Coast Artillery Antiaircraft Gun

Technical Manual

By War Department

©2013 Periscope Film LLC
All Rights Reserved
ISBN#978-1-940453-20-0
www.PeriscopeFilm.com

DISCLAIMER:

This manual is sold for historic research purposes only, as an entertainment. It contains obsolete information and is not intended to be used as part of an actual operation or maintenance training program. No book can substitute for proper training by an authorized instructor.

TECHNICAL MANUAL }
No. 4–325 }

WAR DEPARTMENT,
WASHINGTON, December 16, 1941.

COAST ARTILLERY GUNNERS' INSTRUCTION, ANTIAIRCRAFT GUN, AUTOMATIC WEAPON, AND HEADQUARTERS BATTERIES, FIRST AND SECOND CLASS GUNNERS

Prepared under direction of the
Chief of Coast Artillery

CHAPTER 1

GENERAL

1. **Purpose and scope.**—*a. Purpose.*—This manual is designed primarily for use by organization commanders in the instruction of enlisted men of antiaircraft artillery units (except searchlight batteries) of the Coast Artillery Corps. It may be used also by officers conducting examinations of enlisted men for qualification as gunners, as contemplated by FM 4–150. The questions and answers in the manual are intended merely as a guide and should be supplemented by the extensive use of other questions and answers and by practical demonstrations.

b. Scope.—The topics included in this manual are those prescribed in FM 4–150 for qualification of enlisted men as first and second class gunners in antiaircraft artillery units (except searchlight batteries).

2. Assignment of topics.—The following is the general assignment of topics. Each organization should omit those portions of the assigned chapters. sections, and paragraphs that do not pertain to the particular equipment in use by the organization.

SECOND CLASS GUNNERS

Subjects	105-mm gun batteries	3-inch gun batteries	37-mm gun batteries	Machine-gun batteries	Headquarters batteries, regiments (except supply platoons), or brigades	Supply platoons; regiments, and separate battalions	Headquarters batteries and ammunition trains, battalion
Service of the piece (ch. 2)	Par. 3	Par. 4	Par. 5	Par. 6	Appropriate paragraph.¹	Appropriate paragraph.¹	Appropriate paragraph.¹
Gun and mount (ch. 3)	Sec. I	Sec. II	Sec. III	Sec. IV	Appropriate section.¹	Appropriate section.¹	Appropriate section.¹
Ammunition, fuzes, and projectiles, to include precautions in handling (guns) (ch. 4).	do	Sec. I			do¹	do¹	Do.¹
Ammunition, characteristics, and service (automatic weapons) (ch. 4).			Sec. II	Sec. II			
Motor transportation (ch. 9).²	Pars. 61, 62	Pars. 61, 62	Pars. 61, 62	Pars. 61, 62	Pars. 61, 62	Pars. 61, 62	Pars. 61, 62
Supplies (ch. 11)						Pars. 76–79, incl.	
Nomenclature, action, and maintenance of small arms with which organization is equipped and its ammunition (ch. 12).	Sec. I	Sec. I	Sec. I	Sec. I	Sec. I	Sec. I	Sec. I
Cordage and mechanical maneuvers (ch. 12)	Sec. II	Sec. II	Sec. II	Sec. II	Sec. II	Sec. II	Sec. II

¹ The paragraph or section assigned to a particular supply platoon, to a particular regimental or brigade headquarters battery, or to a particular battalion headquarters battery and ammunition train is determined by the armament of the battalion of which it is a part.

² For organizations having no transportation, motor transportation may be omitted.

4

FIRST CLASS GUNNERS

Subjects	105-mm gun batteries	3-inch gun batteries	37-mm gun batteries	Machine-gun batteries	Headquarters batteries, regiments (except supply platoons) or brigades	Supply platoons, regiments, and separate battalions	Headquarters batteries and ammunition trains, battalion (except ammunition trains)	Ammunition trains, headquarters batteries, and ammunition trains, battalion
Duties of range section (guns) (ch. 5)	Sec. I	Sec. I						
Position-finding apparatus (guns) (ch. 5)	Sec. II	Sec. II						
Pointing and pointing tests (guns) (ch. 5)	Sec. III	Sec. III						
Use, orientation, and adjustment of fire-control instruments (automatic weapons) (ch. 6)			Sec. I	Sec. I				
Determination of firing data (automatic weapons) (ch. 6)			Sec. II	Sec. II				
Conduct and control of fire (automatic weapons) (ch. 6)			Sec. III	Sec. III				
Transporting, handling, and storing ammunition (ch. 4)								Sec. III
Use, orientation, and adjustment of observation instruments (ch. 7)					Sec. I		Sec. I	
Orientation and reconnaissance (ch. 7)					Sec. II		Sec. II	
Elementary definitions for antiaircraft artillery (ch. 8)	Sec. I	Sec. I	Sec. I	Sec. I	Sec. I	Sec. I	Sec. I	Sec. I
Particular definitions pertaining to supplies and supply functions (ch. 8)						Pars. 58, 60		
Particular definitions pertaining to ammunition supply (ch. 8)								Pars. 58, 59
Motor transportation (ch. 9)*	Pars. 63-66, incl.	Pars. 63-66, incl.	Pars. 63-66, incl.	Pars. 63-66, incl.	Pars. 63, 64	Pars. 63-66, incl.	Pars. 63, 64	Pars. 63-67, incl.
Use and care of telephones (ch. 10)	Sec. I	Sec. I	Sec. I	Sec. I	Sec. I	Sec. I	Sec. I	
General knowledge of installation and operation of regimental or battalion telephone system and net, to include the duties of switchboard operators; or Radio communication (ch. 10)					Sec. II or III		Sec. II or III	
Map reading (ch. 12)	Sec. IV	Sec. IV	Sec. IV	Sec. IV	Sec. IV	Sec. III	Sec. IV	Sec. III
Indication, identification, and characteristic features of classes of aircraft (ch. 12)						Sec. IV		Sec. IV

* For organizations having no transportation, motor transportation may be omitted.

CHAPTER 2

SERVICE OF THE PIECE

3. Drill of 105-mm gun section.—*Q.* Normally, how many gun sections are included in a battery of 105-mm guns? *A.* Three gun sections.

Q. How is the gun section formed? *A.* See figure 1.

Q. What are the posts of the gun section? *A.* See figure 2.

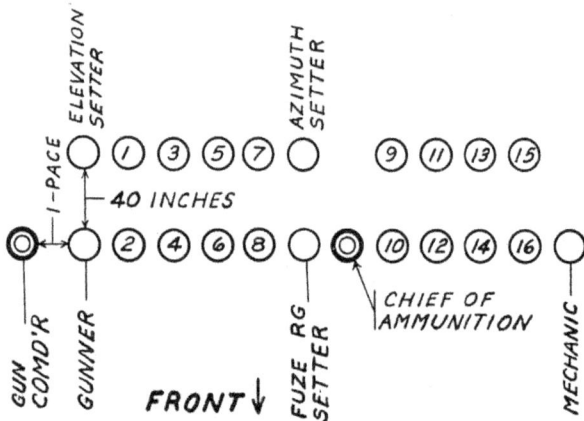

FIGURE 1.—Formation for 105-mm gun section.

Q. What are the duties of the gun commander? *A.* The gun commander is in charge of the gun section and is also chief of the gun squad. He is responsible to the battery executive for—

(1) Training and efficiency of the personnel of his section.

(2) Condition, care, and preparation for action of all matériel, including ammunition, under his charge.

(3) Observance of all safety precautions pertaining to the service of the piece and the handling of ammunition.

(4) Police of the emplacement pertaining to his section.

(5) Care and disposal of empty cartridge cases.

(6) Keeping a record of the number of rounds fired by his gun during a practice or action and of all other data necessary to keep the gun book accurate and up to date.

Q. What are the duties of the chief of ammunition? *A.* The chief of ammunition is in charge of the ammunition squad. He is responsible to the gun commander for—

(1) Training and efficiency of the personnel under his charge.

(2) Condition and serviceability of the ammunition pertaining to the gun.

(3) Observance of all safety precautions in the care and service of the ammunition.

(4) Correct recording of required ammunition data.

(5) Cleaning and disposition of empty cartridge cases.

(6) Uninterrupted service of ammunition to the gun emplacement during the course of a practice or action.

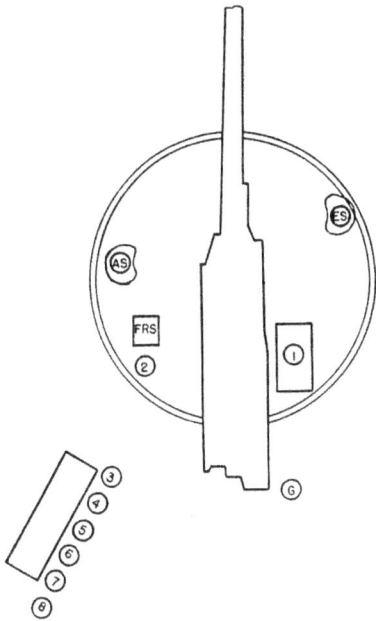

G—Gunner.
AS—Azimuth setter.

ES—Elevation setter.
FRS—Fuze range setter.

FIGURE 2.—Posts of 105-mm gun section.

Q. What are the duties of the artillery mechanic? *A.* The artillery mechanic, assisted by other members of the gun section, makes such minor repairs as can be made with the means at hand. He is responsible for the proper lubrication of and the serviceability and issue of supplies and tools for, the gun and mount. He maintains a lubrication record and also an inventory of all tools, spare parts, and other supplies and equipment under his charge.

Q. What are the duties of the members of the gun section at the various commands? *A.* See table I.

Table I.—*Drill for service of the piece, 105-mm antiaircraft gun*

Details	DETAILS, POSTS	(a) EXAMINE GUN (b) REPORT	TARGET	COMMENCE FIRING (LOAD)	(a) SUSPEND FIRING (b) CEASE FIRING
Gunner...	Procures wiper of cotton waste and can of oil; places wiper and oil in convenient place, and takes post to right rear of piece opposite to and facing end of loading tray.	(a) Assisted by Nos. 1 and 3, examines, cleans, and oils breechblock, breech mechanism, and rammer mechanism; tests firing and rammer mechanisms; examines chamber and bore; and, if necessary, calls upon Nos. 2 to 8, inclusive, for assistance in sponging and cleaning chamber and bore. Assisted by No. 1, inspects oil level in recoil cylinder and front and rear buffer cylinders, under direction of gun commander, and, when necessary, fills those cylinders with recoil oil. When so directed by gun commander, obtains hand air pump and replenishes air pressure in rammer cylinder, being assisted by No. 1. (b) Reports to gun commander, "Breech in order," or reports any defects he is unable to remedy without delay.	Supervises opening of breech and retraction of rammer, holding up end of rammer retracting cam while No. 1 retracts rammer. *When rounds are being placed in loading tray broadside* (at high angles of elevation), as soon as rammer has been retracted, pulls back on rammer head release lever, dropping rammer head to its loading position. When ejected cartridge cases are to be stopped in loading tray, he turns rammer tripping cam to its "engaged" position so that, during last portion of movement of rammer body to rear, rammer head will be forced down into its lower position.	When rounds are being placed in loading tray from rear (at low or medium angles of elevation), drops rammer head after round has been placed in loading tray in front of rammer head. At the command RAM, given by No. 3 (4, 5, 6, 7, or 8) after round has slipped back or been pushed back into contact with a rammer head (at any angle of elevation), he trips rammer by pulling back on rammer hatch release lever and immediately grasps lanyard in his right hand. As soon as breech has closed, he fires piece by a pull on lanyard. In case of misfire he calls, "Misfire," to gun commander and keeps all cannoneers clear of breech.	(a) If an unfired round is in gun, directs No. 1 to retract and drop rammer head just to rear of breech and then drop breechblock with hand lever. After extractors have thrown round back against rammer head, directs No. 1 to continue retracting rammer until it is latched back in loading position, and directs a member of loading detail to remove round from loading tray. (b) Same as SUSPEND FIRING.
Fuze range setter.	Takes post seated on left of gun facing fuze setter.	(a) Examines, cleans, and oils (when necessary) fuze setter, fuze range indicator, and connections thereto, assisted by No. 2. Makes sure that swing bolt nuts are both tight. Releases setting handwheel by means of knob at rear of fuze setter, pulling knob directly to rear. As soon as No. 2 has operated unloaded fuze setter through one or more cycles, tests operation of fuze setter by setting and checking fuze at several values, assisted by Nos. 2 and 4.	Turns adjusting handwheel so as to match mechanical pointer with electrical pointer of fuze range indicator, and thereafter keeps pointers matched.	Continues to keep pointers matched.	(a) Continues to keep pointers matched. (b) Sets mechanical pointer to fuze setting "15" (for mechanical fuze M2) and directs a member of loading detail to remove round in setter. Supervises setting to 15 of any other rounds necessary, checking each fuze as it is removed from setter. Remains at post.

Position					
		(b) Reports to gun commander, "Fuze setter in order," or reports defects he is unable to remedy without delay.			
Azimuth setter.	Takes post seated on left side of gun facing azimuth indicator.	(a) Examines traversing mechanism and assures himself that power is turned on in azimuth indicator. Assists gun commander in checking adjustment of mechanical pointer and synchronization of electrical pointer. (b) Reports to gun commander, "Traversing in order," or reports any defects he is unable to remedy without delay.	Traverses gun so as to match mechanical pointers with electrical pointers of azimuth indicator, and thereafter keeps those pointers matched.	do	(a) Continues to keep pointers matched. (b) Stops matching pointers but remains at post.
Elevation setter.	Takes post seated on right side of gun facing elevation indicator.	(a) Examines elevating mechanism and assures himself that power is turned on in elevation indicator. Assists gun commander in checking adjustment of mechanical pointer and synchronization of electrical pointer. (b) Reports to gun commander, "Elevating in order," or reports any defects he is unable to remedy without delay.	Elevates or depresses gun so as to match mechanical pointers with electrical pointers of elevation indicator, and thereafter keeps those pointers matched.	do	(a) Continues to keep pointers matched. (b) Stops matching pointers but remains at post.
No. 1 (breech operator).	Removes and folds up gun cover, assisted by No. 6, and deposits it at designated place. Takes post on right of gun standing on breech-operating platform facing breech-operating handle.	(a) Assists gunner in examining, cleaning, and oiling breech mechanism, rammer, firing mechanism, chamber, and bore of gun. Assists gunner in inspecting oil level in recoil cylinder and front and rear buffer cylinders, and, when necessary, assists in filling those cylinders with recoil oil. When so directed by gun commander, assists gunner in obtaining hand air pump and replenishing air pressure in rammer cylinder. Sets rammer release tripping cam as directed by gun commander. (b) No duties.	Opens breech by bearing down on operating handle until breechblock is locked open, and immediately raises handle to its vertical position. By means of rammer retracting wheel, retracts rammer to its loading position as directed by gunner.	No duties, unless hand operation of breech is ordered, in which case, opens breech after each round.	(a) If unfired round is in gun, retracts rammer so that head will drop directly behind breech, drops rammer head, opens breech, and eases unfired round back into loading tray by retracting rammer head until rammer is latched back in loading position, as directed by gunner. (b) Same as SUSPEND FIRING.

9

Table I.—*Drill for service of the piece, 105-mm antiaircraft gun*—Continued

Details	DETAILS, POSTS	(a) EXAMINE GUN (b) REPORT	TARGET	COMMENCE FIRING (LOAD)	(a) SUSPEND FIRING (b) CEASE FIRING
No. 2 (fuze setter operator).	Removes muzzle cover and puts it in designated place. Takes post standing to right of fuze range setter facing fuze setter.	(a) Assists fuze range setter in examining, cleaning, and oiling fuze setting mechanism. When setting handwheel has been released, operates unloaded fuze setter through one or more setting cycles (by turning handwheel clockwise) to see that handwheel stops and locks at end of fourth turn of each cycle. Assists fuze range setter in setting and checking fuze at several values. If necessary, assists gunner in sponging bore. (b) No duties.	Locks setting handwheel by turning it clockwise, until it is stopped automatically.	With a round in the fuze setter, rotates setting handwheel quickly until it brings up against stop, turning handle in clockwise direction. On completion of fuze setting operation, calls "Cut." Continues to set fuzes as quickly as projectiles are inserted in fuze setter.	(a) Sets fuze if unset round is in fuze setter or if round is placed in fuze setter after receipt of command. (b) Remains at post prepared to set such fuzes back to 15 as may be necessary.
Nos. 3, 4, 5, 6, 7, and 8 loading detail.	No. 3 procures chamber and bore sponge and wiper of cotton waste, disposes of sponges in designated place, and takes post at the ammunition rack facing breech of gun. No. 4, assisted by Nos. 5, 6, 7, and 8, uncovers ammunition on rack. No. 6 assists No. 1 in removing and disposing of gun cover. Nos. 4, 5, 6, 7, and 8 take	(a) No. 3 assists gunner in examining, cleaning, and oiling breech mechanism, rammer, and bore of gun. Sets breech operating cam plunger for automatic or hand operation of breech, as directed by gun commander. No. 4, assisted by Nos. 5, 6, 7, and 8, checks ammunition on rack. When called upon by gunner, all six men assist in sponging chamber and bore of gun. No. 4 assists fuze range setter in testing operation of fuze setter by setting and checking fuze at several values. (b) No. 4 reports to gun commander, "Ammunition handling in order," or reports such defects as he is unable to remedy without delay.	No. 3 takes round from ammunition rack, right hand grasping base of cartridge case, with rotating band of projectile resting in crook of left arm, and proceeds by most direct route to fuze setter. Rests projectile on edge of mouth of fuze setter and then inserts round into setter with right hand with one continuous motion and with sufficient force to complete seating. When round is locked in fuze setter, releases round immediately. Nos. 4, 5, 6, 7, and 8 grasp round as prescribed for No. 3 and proceed to left of fuze setter, where they stand in readi-	As soon as No. 2 has called "Cut," No. 3 grasps base of cartridge case with both hands and starts round to rear with quick pull *without rotation*. When firing is at high angles of elevation, catches rotating band of projectile in crook of his left arm, with right hand remaining under base end of case, steps toward gun, and deposits round, broadside, in loading tray with right hand just clearing rammer head. When firing is at low or medium angles of elevation, when loading tray is too high for broadside loading, catches rotating band of projectile in palm of left hand, with right hand remaining under base end of case, steps toward gun, inserts nose of round in rear of	(a) If unfired round is in gun, as soon as it is ejected onto loading tray, it is removed and laid aside by that member of loading detail who loaded it. If there is no round in fuze setter, the next succeeding member of loading detail inserts one, and remaining members of loading detail stand ready to continue service of ammunition when firing is resumed. (b) If there is a round in fuze setter, it is removed and returned to ammunition rack by member of loading detail as soon as it

posts at ammunition rack, in column behind No. 3.		ness to deposit their rounds in turn, in setter.	loading tray, and pushes round forward until base of case is in front of rammer head. In either method of loading, after rammer head has been dropped, adjusts round backward into contact with rammer head, steps away quickly, and commands: RAM. Returns to ammunition rack, obtains another round, and falls in behind No. 8 on left of fuze setter. Nos. 4, 5, 6, 7, and 8, in turn, perform operations just described for No. 3 as firing proceeds.	has been set to 16. Other rounds which have been laid aside with their fuzes set are inserted in fuze setter by members of loading detail; when fuzes are set to 15, rounds are returned to ammunition rack.
Nos. 9 to 16, inclusive (ammunition squad). No. 9 obtains pair of asbestos gloves and puts them in convenient place near breech of gun. Nos. 9 to 16, inclusive, are posted by chief of ammunition squad in such manner as to expedite supply of ammunition.	(a) Remove from boxes or crates and prepare necessary ammunition for contemplated practice or action. Fill ammunition racks at gun emplacement. (b) No duties.	No. 9 stands ready to receive more empty cartridge cases from emplacement. The remaining members of ammunition squad stand ready to keep rack at gun emplacement filled with ammunition as it is used.	No. 9 catches (or takes from loading tray) empty cartridge cases as they are ejected from gun and removes them from emplacement. Remaining members of squad continue supply of ammunition to rack. Whenever practicable without interfering with gun squad, or during lulls in firing, collect and place at designated point all empty cases preparatory to decapping and cleaning for salvage.	(a) Collect all empty cases, fill ammunition cases, and return to their posts. (b) Same as SUSPEND FIRING.

4. Drill of 3-inch gun section.—Instruction in the service of the piece should be practical and should include the drill and emplacement of the gun.

FRONT   I PACE

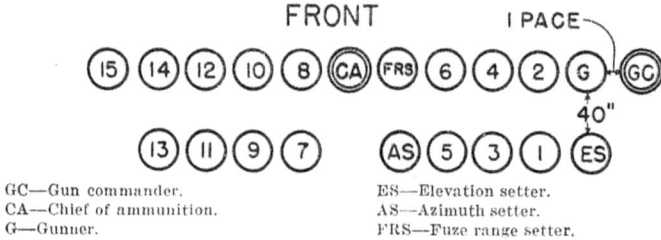

GC—Gun commander.
CA—Chief of ammunition.
G—Gunner.

ES—Elevation setter.
AS—Azimuth setter.
FRS—Fuze range setter.

FIGURE 3.—Formation for 3-inch gun section.

Q. How is the gun section formed? *A.* See figure 3.

Q. What are the posts of the gun section? *A.* See figure 4.

Q. What is the reason for the commands EXAMINE GUN and REPORT? *A.* To check the operating condition of the gun. At the command

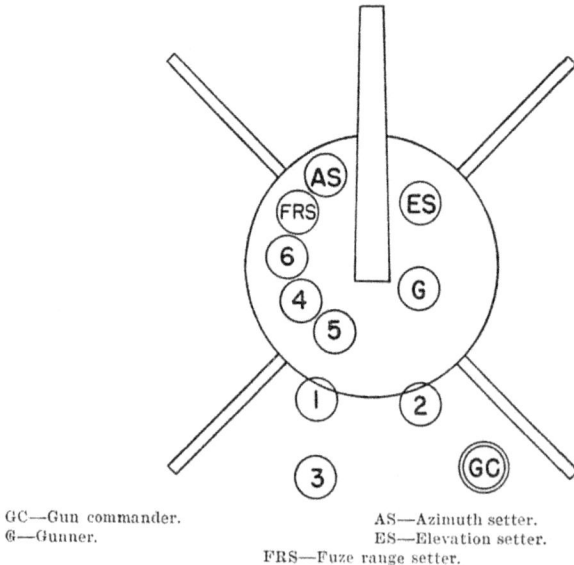

GC—Gun commander.
G—Gunner.

AS—Azimuth setter.
ES—Elevation setter.

FRS—Fuze range setter.

FIGURE 4.—Posts for 3-inch gun section.

EXAMINE GUN, certain members of the gun section test the functioning of various parts of the gun and mount. At the command REPORT, these individuals tell the gun commander whether their particular parts of

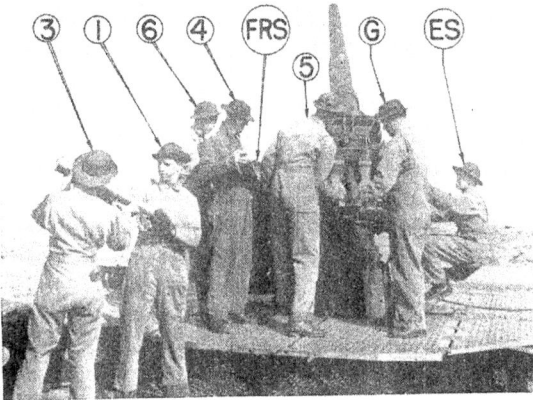

FIGURE 5.—Round being rammed, M3 gun on M2A2 mount.

the gun are in order, whether they require minor adjustment, or whether there are defects which cannot be corrected immediately.

Q. What are the duties of the members of the gun section at the various commands? *A.* See tables II, III, and IV, and figures 5, 6, and 7.

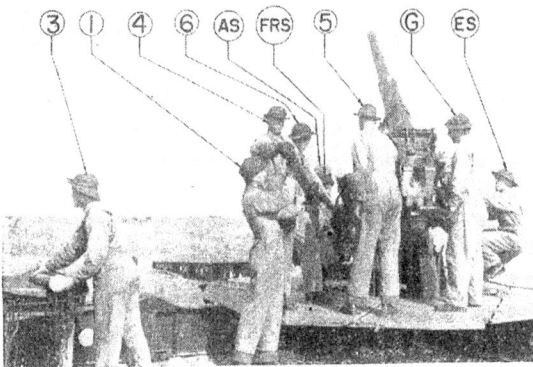

FIGURE 6.—Round leaving continuous fuze setter M8.

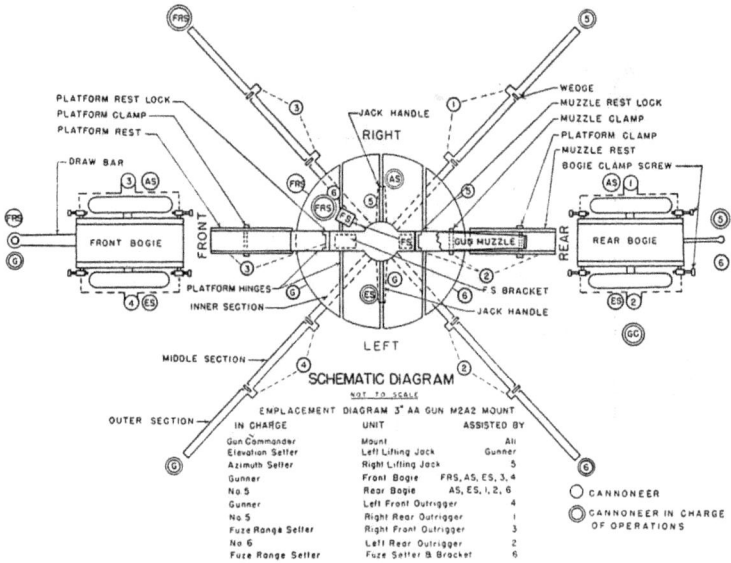

FIGURE 7.—Emplacement diagram, 3-inch gun on M2A2 mount.

TABLE II.—*Drill for service of the piece, 3-inch AA gun with automatic data transmission system and M8 fuze setter*

[3-inch AA gun M3 on mobile mounts M2A1 and M2A2 and 3-inch AA gun M4 on fixed mount M3A1]

Details	DETAILS, POSTS	(a) EXAMINE GUN (b) REPORT	TARGET	COMMENCE FIRING	(a) SUSPEND FIRING (b) CEASE FIRING
Gunner	The gunner procures a wiper of cotton waste and a can of oil, he places the wiper and oil in convenient place, removes the breech cover, assisted by No. 5, disposes of it at the designated place, and takes post to the right rear of the gun opposite and facing the breech.	(a) Assisted by No. 5, he examines, cleans, and oils the breechblock and breech mechanism, tests the firing mechanism, examines the chamber and bore and, if necessary, calls upon Nos. 1, 2, 3, and 5 for assistance in sponging and cleaning the chamber and bore. (b) He reports to the gun commander, "Breech in order," or reports any defects he is unable to remedy without delay.	He opens the breech by bearing down on the operating handle until the breechblock is locked open and immediately raises the operating handle to its vertical position. He assumes a position facing the breech convenient for loading yet clear of the recoil and with his feet well braced and the lanyard grasped in his right hand.	As soon as the first round is placed in loading position by No. 5 he rams the cartridge. As soon as the closing breech has knocked his left hand clear he fires the gun by pulling the lanyard with his right hand. He continues ramming and firing without shifting position or releasing the lanyard. In case of misfire he calls, "Misfire," to the gun commander and keeps all cannoneers clear of the breech while the prescribed safety precautions are taken.	(a) If an unfired round is in the gun, he unloads it by bearing down on the operating handle as soon as No. 5 is ready to receive the cartridge, and remains at his post. (b) Same as SUSPEND FIRING.
Fuze range setter (fuze setter).	He takes post seated on the fuze setter's seat, facing the fuze setter.	(a) Assisted by No. 6, he examines, cleans, and oils (where necessary) the fuze setter, fuze range indicator, and connections thereto. Assisted by No. 4, he tests the operation by setting and checking a fuze at several values. (b) He reports to the gun commander, "Fuze setter in order," or reports such defects as he is unable to remedy without delay.	He turns the fuze range handwheel so as to match the mechanical pointer with the electrical pointer of the fuze range indicator and thereafter keeps them matched.	He continues to keep his pointers matched.	(a) He continues to keep his pointers matched. (b) He turns his mechanical pointer to "safe" and directs No. 4 to remove the round from the setter. He supervises the setting at "safe" of any other rounds necessary, checking each fuze as it is removed from his setter. Thereafter he remains at his post.

TABLE II.—*Drill for service of the piece, 8-inch AA gun with automatic data transmission system and M8 fuze setter*—Continued

Details	DETAILS, POSTS	(a) EXAMINE GUN (b) REPORT	TARGET	COMMENCE FIRING	(a) SUSPEND FIRING (b) CEASE FIRING
Azimuth setter.	He takes post seated on the left side of the gun, facing the azimuth indicator.	(a) He examines the traversing mechanism and azimuth indicator and connections. (b) He reports to the gun commander, "Traversing in order," or reports any defects he is unable to remedy without delay.	He traverses the gun so as to match the mechanical pointers with the electrical pointers of the indicator and thereafter keeps those pointers matched.	He continues to keep his pointers matched.	(a) He continues to keep his pointers matched. (b) He stops matching his pointers but remains at his post.
Elevation setter.	He takes post seated on the right side of the gun, facing the elevation indicator.	(a) He examines the elevating mechanism and elevation indicator and the connections thereto. Should a difficulty in elevating indicate the need of a temperature adjustment to the equilibrator, he reports that fact to the gun commander at once. (b) He reports to the gun commander, "Elevating in order," or reports any defects he is unable to remedy without delay.	He elevates or depresses the gun so as to match the mechanical pointers with the electrical pointers of the elevation indicator and thereafter keeps those pointers matched.	do	(a) He continues to keep his pointers matched. (b) He stops matching his pointers but remains at his post.
Ammunition handlers, Nos. 1, 2, 3, and 4.	No. 4 removes, then folds up the gun cover, assisted by No. 2, deposits it at the designated place, and takes post immediately in rear of the fuze setter, facing it, standing on the gunner's platform. No. 1, assisted by No. 3, uncovers the ammunition and takes post on the ground about 3 feet behind No. 4. No. 3 takes post at the stack	(a) No. 4 inserts a round in the fuze setter, when called for by No. 6, for the purpose of testing the fuze setter, and examines the ammunition, assisted by Nos. 1 and 3. Nos. 1 and 3 assist No. 4 in inspecting and arranging the ammunition near the gun. If necessary, they assist the gunner in cleaning and sponging the bore. No. 2 assists the gunner, if necessary, in cleaning and sponging the bore, and examines the emplacement, removing and disposing of any obstructions and covering soft spots on the ground that might interfere	No. 4 takes a round from No. 1 and inserts it in the fuze setter, presses down on the base end of the case with his right hand, brings his left hand over and strikes a quick slap against the release lever on top side of fuze setter. After No. 6 has completed one turn of the setting crank No. 4 removes his right hand pressure from the base of the projectile, turns to receive another round from No. 1, and stands	No. 4 continues loading the fuze setter as fast as rounds are removed therefrom by No. 5, receiving the rounds from No. 1. Nos 1 and 3 continue serving ammunition to No. 4. No. 2 clears the empty cases from the emplacement.	(a) If no round is in the fuze setter, No. 4 inserts one and stands ready to continue loading the fuze setter. Nos. 1 and 3 stand ready to serve ammunition to No. 4. No. 2, if an unfired round remains in the gun, receives it from No. 5, lays it aside, and continues clearing the empty cases from the emplacement. (b) Nos. 1, 3, and 4 procure and assist in setting at "safe" such rounds as have been set and remain unfired, return

	of ammunition. No. 2 assists No. 4 in removing and folding up the gun cover and takes post to the rear of the gunner, standing on the ground and facing the breech, just clear of the gunner's platform.	with the smooth functioning of Nos. 1, 3, and 4. (b) No. 4 reports to the gun commander, "Ammunition in order," or reports any defects he is unable to remedy without delay.	ready to load the fuze setter again. No. 1 takes a round from No. 3 and passes it to No. 4. He repeats the operation. No. 3 takes a round from the stack and passes it to No. 1. He repeats the operation. No. 2 has no duties. (For target practice see note 1.)	He removes a round from the fuze setter and holds it in the loading position at the breech. As soon as the round is loaded (by the gunner) he repeats the operation.	them to the stack, and remain at their posts. No. 2 continues clearing the empty cases from the emplacement. When the emplacement is cleared he remains at his post.
Relayer, No. 5.	No. 5 procures the chamber and bore sponges and a wiper of cotton waste, disposes of the sponges in the designated place, assists the gunner in removing the breech cover, and takes post about 2 feet to the left and rear of the breech, facing the breech.	(a) He assists the gunner in examining, cleaning, and oiling the breech mechanism, firing mechanism, chamber, and bore of the gun. He is responsible that the breech operating cam plunger is set for automatic or hand operation of the breech as directed by the gun commander. (b) No duties.	No duties.----------		(a) If an unfired round is in the gun, he places his right palm behind the breechblock at the signal of the gunner, catches the ejected round and passes it to No. 2, and remains at his post. (b) Same as SUSPEND FIRING.
Fuze setter operator, No. 6.	No. 6 removes the muzzle cover and deposits it at the designated place. He takes post standing at the right rear of the fuze range setter, facing the fuze setter.	(a) He assists the fuze range setter in examining, cleaning, and oiling the fuze setting mechanism. Checks to see proper rings are in fuze setter. (b) No duties.	He places his hand on handle so that as soon as the release lever is struck by No. 4, the crank will start. He turns the handle until it comes in contact with the stop at which time he calls "Cut" to signify that the fuze is set and the round can be withdrawn. (For target practice see note 1.)	He continues setting fuzes as quickly as projectiles are inserted in the fuze setter, calling "Cut" as each fuze is set.	(a) He sets the fuze if an unset round is in the fuze setter or if a round is placed in the fuze setter after receipt of the command. (b) He remains at his post prepared to set such fuzes back to "safe" as may be necessary.

17

TABLE II.—*Drill for service of the piece, 3-inch AA gun with automatic data transmission system and M8 fuze setter*—Continued

Details	DETAILS, POSTS	(a) EXAMINE GUN (b) REPORT	TARGET	COMMENCE FIRING	(a) SUSPEND FIRING (b) CEASE FIRING
Ammunition squad, Nos. 7 to 15, inclusive. (See note 2.)	Nos. 7 to 15, inclusive, are posted by the chief of ammunition in such manner as to expedite the supply of ammunition and perform such other duties as may be directed.	(a) Nos. 7 to 15, inclusive, under the direction of the chief of ammunition, remove from the boxes or crates and prepare the necessary ammunition for the contemplated practice or action. If practice or action is imminent, they place an ammunition supply in a stack at the gun emplacement. (b) No duties.	Nos. 7 to 15, inclusive, stand ready to replenish the ammunition supply at the gun position.	They continue the supply of ammunition to the gun position.	(a) After collecting all empty cases and replenishing the ammunition supply at the gun, they remain at their posts unless otherwise directed. (b) Same as SUSPEND FIRING.

NOTES

1. The service of the piece, as written, prescribes that the continuous fuze setter be loaded at the command TARGET. This is logical for all service firings; however, in target practice it is not desirable that the fuze setter be loaded and the fuze set until the target has reached approximately safe firing conditions. Therefore, for target practice, Nos. 4 and 6 should have no duties at the command TARGET, but the battery commander should give a warning command STAND BY as the safe field is approached, at which time No. 4 loads the fuze setter, No. 6 sets the fuze as prescribed herein, and No. 5 awaits the command COMMENCE FIRING before removing the round from the setter.

2. This table is for the organization of a mobile unit. For semimobile units, the numbers of the ammunition squad are from 7 to 12 as three chauffeurs are not included in the organization for this type unit.

TABLE III.—*Drill for emplacing M8 gun on M2A1 or M2A2 mobile mount*

Details	PREPARE FOR ACTION	JACKS DOWN	JACKS HALT	JACKS UP
Gunner	The gunner is in charge of the jack float channel, the front bogie, and the left front outrigger. Assisted by No. 5, he procures the jack float channel, puts it in place, and centers it under the jack floats.	He assists the elevation setter in raising the mount.	He takes post at the drawbar of the front bogie and removes the bogie assisted by the fuze range setter, the azimuth setter, elevation setter, and Nos. 3 and 4. Assisted by the fuze range setter and Nos. 3 and 4, he unfolds the front outriggers. He then takes post at the outer section of the left front outrigger and, assisted by No.	Assisted by No. 4, he maneuvers the left front outrigger to maintain the mount level while it is being lowered. When the jack operators are clear, he unlocks the left front section of the platform from the platform rest, and lowers and locks it in its firing position. Assisted by No. 4 he fills dirt under the left front outrigger if necessary.

Fuze range setter	The fuze range setter is in charge of the right front outrigger, the fuze setter bracket, and the fuze setter.	He takes post at the drawbar of the front bogie.	4, swings it out to its stop and maneuvers it to assist in the removal of the rear bogie. He assists in removing the front bogie. He assists in unfolding the front outriggers. He then takes post at the right front outrigger and, assisted by No. 3, swings it out to its stop and maneuvers it to assist in the removal of the rear bogie.	Assisted by No. 3, he maneuvers the right front outrigger to maintain the mount level while it is being lowered. When the jack operators are clear, he unlocks the right front section of the platform from the platform rest and lowers and locks it in its firing position. Assisted by No. 6, he moves the fuze setter bracket from its traveling position, installs it in its firing position, unfolds the fuze setter's seat, and installs the fuze setter on its bracket. Assisted by No. 3, he fills dirt under the right front outrigger if necessary.
Azimuth setter	He is in charge of the right lifting jack and the azimuth seat. He runs the right lifting jack down by means of the fast motion handwheel until the float is in contact with the jack float channel and sets the jack lever ratchet for JACKS DOWN.	Assisted by No. 5 he runs the right jack down by means of the jack lever, raising the mount.	He unlocks the azimuth seat from its traveling position and swings it out from the mount. He assists in the removal of the front and rear bogies.	He runs the right jack up until the mount is resting on the ground and continues the operation until the jack is fully retracted. He locks the azimuth seat in its firing position and operates one leveling handle as directed by the gun commander.
Elevation setter	He is in charge of the left lifting jack and the elevation seat. He runs the left lifting jack down by means of the fast motion handwheel until the float is in contact with the jack float channel (which has been put in place by the gunner and No. 5) and sets the jack lever ratchet for JACKS DOWN.	Assisted by the gunner, he runs the left jack down by means of the jack lever, raising the mount.	He unlocks the elevation seat from its traveling position and swings it out from the mount. He assists in the removal of the front and rear bogies.	He runs the left jack up until the mount rests on the ground and continues the operation until the jack is fully retracted. He locks the elevation seat in its firing position and elevates the gun as necessary to free the muzzle from its clamp. He operates one leveling handle as directed by the gun commander.
Ammunition handler, No. 1	No duties..........	He takes post at the right wheel of the rear bogie.	After the front outriggers have been extended and wedged, No. 1 unlocks the two right-hand clamp screws of the rear bogie and the right rear outrigger clamp screws. He assists in removing	He assists No. 5 in maneuvering the right rear outrigger while the mount is being lowered to the ground. He assists No. 5 in filling dirt under the right rear outrigger if necessary.

TABLE III.—*Drill for emplacing M3 gun on M2A1 or M2A2 mobile mount*—Continued

Details	PREPARE FOR ACTION	JACKS DOWN	JACKS HALT	JACKS UP
Ammunition handler No. 1.—Con.			the rear bogie. He assists in unfolding the rear outriggers. He inserts the wedges in the right rear outrigger and assists No. 5 in swinging the outrigger to its stop. Uncouples electric brake connections on rear bogie assisted by No. 2.	
Ammunition handler, No. 2.	No duties.	He takes post at the left wheel of the rear bogie.	After the front outriggers have been extended and wedged. No. 2 unlocks the two left-hand clamp screws of the rear bogie and the left rear outrigger clamp screws. He assists in removing the rear bogie. He assists in unfolding the rear outriggers. He inserts the wedges in the left rear outrigger and assists No. 6 in swinging the outrigger to its stop. Assists No. 1 in uncoupling electric brake connections on rear bogie.	He assists No. 6 in maneuvering the left rear outrigger while the mount is being lowered to the ground. He unlocks the muzzle clamp and directs the elevation setter to elevate the gun. He unlocks the muzzle rest and lays it out on the ground. He then assists No. 6 in filling dirt under the left rear outrigger if necessary.
Ammunition handler, No. 3.	No. 3 unlocks the two right-hand clamp screws of the front bogie and the right front outrigger clamp screws. Uncouples electric brakes on front bogie assisted by No. 4.	He takes post at the right wheel of the front bogie.	He assists in removing the front bogie. He assists in unfolding the front outriggers. He inserts the wedges in the right front outrigger and assists the fuze range setter in swinging the outrigger to its stop and maneuvering it.	He assists the fuze range setter in maneuvering the right front outrigger while the mount is being lowered to the ground. He then unlocks the platform rest and lays it out on the ground. He assists the fuze range setter in filling dirt under the right front outrigger if necessary.
Ammunition handler, No. 4.	No. 4 unlocks the two left-hand clamp screws of the front bogie and the left front outrigger clamp screws. Assists No. 3 in uncoupling electric brake connections.	He takes post at the left wheel of the front bogie.	He assists in removing the front bogie. He assists in unfolding the front outriggers. He inserts the wedges in the left front outrigger and assists the gunner in swinging the outrigger to its stop and maneuvering it.	He assists the gunner in maneuvering the left front outrigger while the mount is being lowered to the ground. He then assists the gunner in filling dirt under the left front outrigger if necessary.
Relayer, No. 5.	No. 5 is in charge of the rear bogie and the right rear outrigger. He assists the gunner with the jack float channel.	He assists the azimuth setter in raising the mount.	He takes post at the drawbar of the rear bogie and, after the front bogie is removed and the outriggers extended, removes the rear bogie, assisted by the	Assisted by No. 1 he maneuvers the right rear outrigger to maintain the mount level while it is being lowered. When the jack operators are clear he un-

Fuze setter operator, No. 6.	No. 6 is in charge of the left rear outrigger.	He takes post at the drawbar of the rear bogie.	azimuth setter, elevation setter, and Nos. 1, 2, and 6. Assisted by Nos. 1, 2, and 6, he unfolds the rear outriggers. He then takes post at the outer section of the right rear outrigger and, assisted by No. 1, swings it out to its stop. He assists in removing the rear bogie. He assists in unfolding the rear outriggers. He then takes post at the outer section of the left rear outrigger and, assisted by No. 2, swings it out to its stop.	locks the right rear section of the platform from the muzzle rest and lowers and locks it in its firing position. Assisted by No. 1, he fills dirt under the right rear outrigger if necessary. Assisted by No. 2, he maneuvers the left rear outrigger to maintain the mount level while it is being lowered. When the jack operators are clear, he unlocks the left rear section of the platform from the muzzle rest and lowers and locks it in its firing position. He assists the fuze range setter in installing the fuze setter bracket and the fuze setter. Assisted by No. 2, he fills dirt under the left rear outrigger if necessary.
Ammunition squad, Nos. 7 to 15, inclusive. (See note 3.)	Nos. 7 to 15, inclusive, under the supervision of the chief of ammunition, unload and prepare ammunition for service and perform such other duties connected with the preparation of the emplacement as may be directed by the gun commander.	As directed	As directed	As directed.

Caution: Care should be exercised that the rear leveling lever is not moved while the gun is secured to the traveling lock, as such movement will damage the elevating mechanism assemblies.

NOTES

1. The end of the mount equipped with the muzzle rest is the rear of the mount. Due to interference of various members of the mount and bogies, it is not allowable to put the mount in traveling position so that the muzzle of the gun points forward when traveling, although the bogies are interchangeable. The bogie equipped with the lunette is the front bogie and will always be placed in position at the front of the mount (breech end of gun).

2. The mount M2A1, unless modified after issue, is equipped with pneumatic type (air) brakes. References to electric brake connections should be changed in this case to read "air hose connections" when applied to rear bogies and eliminated when applied to front bogies.

3. This table is for the organization of a mobile unit. For semimobile units, the numbers of the ammunition squad are from 7 to 12, as three chauffeurs are not included in the organization of this type unit.

21

TABLE IV.—*Drill for placing M3 gun on M2A1 or M2A2 mobile mount in traveling position*

Details	MARCH ORDER	JACKS DOWN (1)	JACKS HALT (1)	JACKS DOWN (2)	JACKS HALT (2)	JACKS UP
Gunner	Assisted by No. 5 he places the breech cover over the breech. Assisted by the fuze range setter and Nos. 3 and 4, he folds the rear sections of the platform and locks them in traveling position.	He assists the elevation setter in maneuvering the left lifting jack.	Assisted by the fuze range setter, and Nos. 3 and 4, he drives the wedges out of the rear outriggers, folds them, and clamps them in traveling position.	He assists the elevation setter in operating the left lifting jack.	He assists the azimuth setter, the fuze range setter, and Nos. 3, 4, and 6 in maneuvering the rear bogie. Takes post at left wheel of front bogie.	He assists the fuze range setter and Nos. 1, 2, 5, and 6 in replacing the front bogie under gun. After the front bogie is replaced, he replaces the jack float channel in traveling position. No. 5 assists in the operation.
Fuze range setter.	Assisted by No. 6, he removes fuze setter and fuze setter bracket, placing them in traveling position. He assists the gunner and Nos. 3 and 4 to fold the rear sections of the platform and to lock them in position.	He takes post by the rear outriggers.	Assists the gunner and Nos. 3 and 4 in driving the wedges out of the rear outriggers and clamping those outriggers in traveling position.	He takes post at the drawbar of the rear bogie.	He assists the gunner, the azimuth setter, and Nos. 3, 4, and 6 in maneuvering the rear bogie. Takes post at right wheel of front bogie.	He assists the gunner, Nos. 1, 2, 5, and 6 in replacing the front bogie under gun. With the aid of No. 4 he connects the electric brakes on the rear bogie.
Azimuth setter.	He traverses the gun until it is in position directly over the muzzle rest. He unlocks the azimuth seat from firing position.	Assisted by No. 5, he runs the right lifting jack down, as directed by the gun commander, until the outriggers are broken free of the ground.	No duties	Assisted by No. 5, he runs the right lifting jack down, as directed by the gun commander, until the rear bogie may be replaced.	Assisted by the gunner, the fuze range setter, and Nos. 3, 4, and 6, replaces the rear bogie and clamps it in position.	He runs the right lifting jack up as directed by the gun commander to permit the gun to settle on the front bogie. After the bogie is locked in position, he runs the lifting jack up fully and locks it. He locks the azimuth seat in traveling position.
Elevation setter.	He depresses the gun as it is traversed by the azimuth setter until the gun is in position to be clamped to the muzzle rest. He un-	Assisted by the gunner, he runs the left lifting jack down, as directed by the gun com-	do	Assisted by the gunner, he runs the left lifting jack down, as directed by the gun com-	No duties	He runs the left lifting jack up as directed by the gun commander to permit the gun to settle on the front bogie. After the bogie is

(continued)	locks the elevation seat from firing position.	mander, until the outriggers are broken free of the ground.		mander, until the rear bogie may be replaced.		locked in position, he runs the lifting jack up fully and locks it. He locks the elevation seat in traveling position.
Ammunition handler, No. 1.	He places the platform rest in traveling position and, when gun is traversed and depressed properly, clamps the gun to the muzzle rest. Assists Nos. 2, 5, and 6 to fold the front sections of the platform and to lock them in position.	No duties	Takes post at end of right front outrigger.	Bears down on the end of right front outrigger while the jacks are run down enough to replace the rear bogie.	He assists Nos. 2, 5, and 6 with the front outriggers and takes post at right wheel of front bogie.	He assists the fuze range setter, gunner, and Nos. 2, 5, and 6 in replacing front bogie. Tightens bogie clamps. He replaces additional equipment at direction of the gun commander.
Ammunition handler, No. 2.	He assists Nos. 1, 5, and 6 to fold the front sections of the platform and to lock them in position.	do	Takes post at end of left front outrigger.	Bears down on the end of left front outrigger while the jacks are run down enough to replace the rear bogie.	He assists Nos. 1, 5, and 6 with the front outriggers and takes post at the left wheel of the front bogie.	He assists the fuze range setter, gunner, and Nos. 1, 5, and 6 in replacing the front bogie. Assists No. 3 in connecting the electric brakes on the front bogie. Tightens bogie clamps.
Ammunition handler, No. 3.	He assists the gunner, the fuze range setter, and No. 4 to fold the rear sections of the platform and to lock them in position.	He takes post by the rear outriggers.	Assists the gunner, the fuze range setter, and No. 4 in driving the wedges out of the rear outriggers and in folding and clamping those outriggers in traveling position.	He takes post at the right wheel of the rear bogie.	He assists the azimuth setter, gunner, fuze range setter, and Nos. 4 and 6 in maneuvering the rear bogie. Tightens bogie clamps.	Assisted by No. 2, he connects the electric brakes on the front bogie.
Ammunition handler, No. 4.	He places the muzzle rest in traveling position. He assists the gunner, the fuze range setter, and No. 3 to fold the rear sections of the platform and to lock them in position.	do	Assists the gunner, the fuze range setter, and No. 3 in driving the wedges out of the rear outriggers and in folding and clamping those outriggers in traveling position.	He takes post at the left wheel of the rear bogie.	He assists the gunner, the fuze range setter, the azimuth setter, and Nos. 3 and 6 in maneuvering the rear bogie. Tightens bogie clamps.	He assists the fuze range setter in connecting the electric brakes on the rear bogie.

23

TABLE IV.—*Drill for placing M3 gun on M2A1 or M2A2 mobile mount in traveling position*—Continued

Details	MARCH ORDER	JACKS DOWN (1)	JACKS HALT (1)	JACKS DOWN (2)	JACKS HALT (2)	JACKS UP
Relayer, No. 5.	He assists the gunner to place the breech cover over the breech. Assisted by Nos. 1, 2, and 6, he folds the front sections of the platform and locks them in traveling position.	He assists the azimuth setter in maneuvering the right lifting jack.	No duties	He assists the azimuth setter in maneuvering the right lifting jack.	After the rear bogie is replaced, assisted by Nos. 1, 2, and 6, he removes the wedges in the front outriggers and folds and clamps those outriggers in traveling position. He then takes post at the drawbar of the front bogie.	He directs the placing of the front bogie under gun, aided by the fuze range setter, the gunner, and Nos. 1, 2 and 6. He then assists the gunner to replace the jack float channel in traveling position.
Fuze setter operator, No. 6.	He replaces the muzzle cover, and assists the fuze range setter in handling the fuze setter and fuze setter bracket. He assists Nos. 1, 2, and 5 to fold the front sections of the platform and to lock them in traveling position.	No duties	....do....	No duties	He assists the azimuth setter, the gunner, the fuze range setter, and Nos. 3 and 4 in maneuvering the rear bogie. After rear bogie is replaced, assisted by Nos. 1, 2, and 5, he removes the wedges in the front outriggers, and folds and clamps these outriggers in traveling position. He then takes post at drawbar of front bogie.	He assists the fuze range setter, the gunner, and Nos. 1, 2, and 5 to replace the front bogie. He replaces additional equipment at the direction of the gun commander.
Ammunition squad, Nos. 7 to 15, inclusive. (See note 3.)	Nos. 7 to 15, inclusive, under the supervision of the chief of ammunition, perform such duties in salvaging ammunition and quitting the emplacement as may be directed.	As directed	As directed	As directed	As directed	As directed.

Caution: Care should be exercised that the rear leveling lever is not moved while the gun is secured to the traveling lock, as such movement will damage the elevating mechanism assemblies.

NOTES

1. The end of the mount equipped with the muzzle rest is the rear of the mount. Due to interference of various members of the mount and bogies, it is not allowable to put the mount in traveling position so that the muzzle of the gun points forward when traveling, although the bogies are interchangeable. The bogie equipped with the lunette is the front bogie and will always be placed in position at the front of the mount (breech end of gun).

2. The mount M2A1, unless modified after issue, is equipped with pneumatic type (air) brakes. References to electric brake connections should be changed in this case to read "air hose connections" when applied to rear bogies and eliminated when applied to front bogies.

3. This table is for the organization of a mobile unit. For semimobile units, the numbers of the ammunition squad are from 7 to 12, as three chauffeurs are not included in the organization of this type unit.

5. Drill of 37-mm gun section.—*Q.* What is the personnel of a 37-mm gun section? *A.* A gun commander (sergeant), a gunner (corporal), and 6 numbered cannoneers (No. 6 is a chauffeur).

Q. What is the formation for the 37-mm gun section? *A.* See figure 8.

Q. What are the posts of the 37-mm gun section? *A.* See figure 9.

Q. What is the fire unit for 37-mm guns? *A.* The platoon consisting of two 37-mm guns.

Q. What are the duties of the members of the gun section at the various commands? *A.* See tables V and VI.

Q. How is the gun loaded? *A.* At the command TARGET. No. 3 opens the feed box cover, calls "Prime," and holds the carrier up with his hand until it is caught by the carrier catch while No. 2 pulls

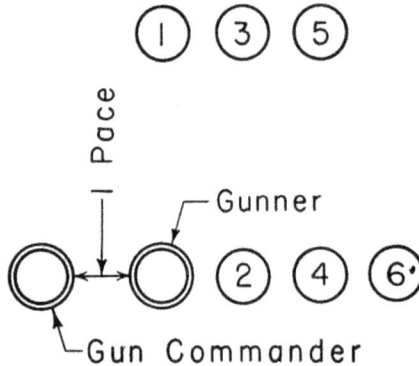

FIGURE 8.—Formation of 37-mm gun section.

the lock frame to the rear and locked position with the priming rod. No. 3 then calls "Ammunition," receives a loaded ammunition clip from No. 4, places it, with the prongs down, on the feedway and pushes it into the receiver until its end is against the carrier catch. He then closes the feed box cover.

At the command LOAD given by the gun commander, No. 3 pushes the ammunition clip into the receiver against the carrier catch until the carrier is tripped, allowing the lock frame to go forward to load the first round.

Q. How is the gun unloaded? *A.* No. 3 folds back the feed box cover. Then No. 2 pulls out on the clip holding pawl button and No. 3 raises the feed pawl lifter button. At the same time No. 3 draws the ammunition clip back and out of the feed box. No. 3 calls "Prime" and No. 2 pulls the lock frame to the rear and locked

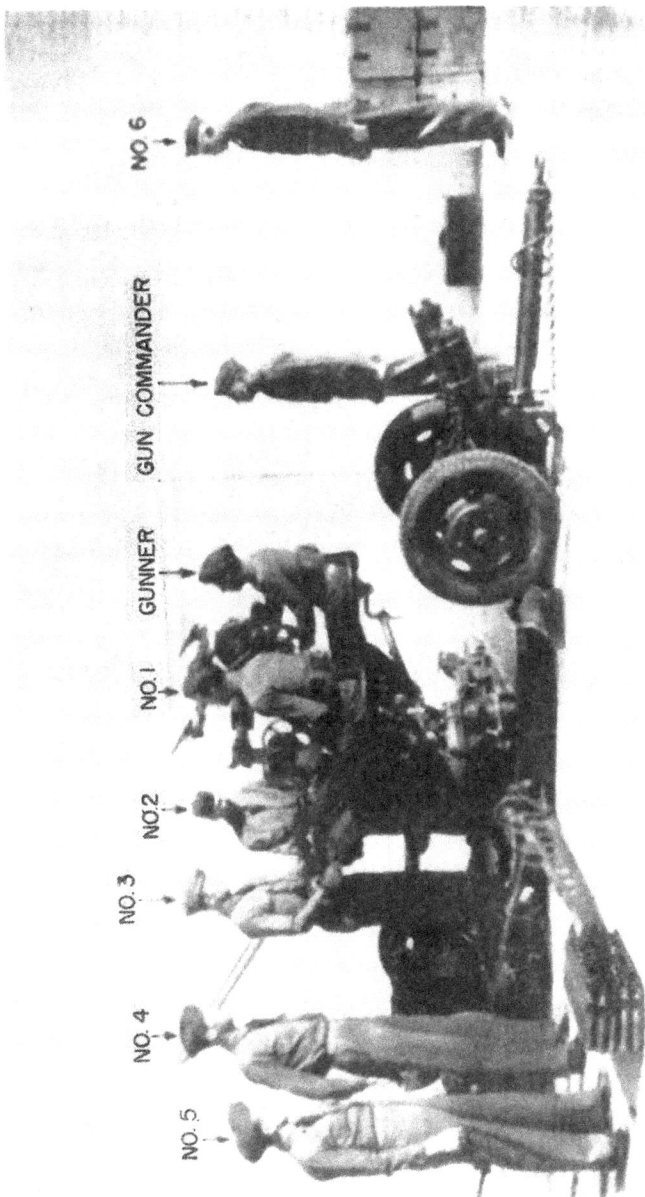

FIGURE 9.—Posts of 37-mm gun section.

position with the priming rod. No. 3 lifts the carrier up during the priming operation until it is held by the carrier catch and catches the unfired round as the lock frame is pulled to the rear. No. 3 then inspects the receiver and breech to see that they are empty and pushes over the carrier catch, allowing the lock frame to go forward.

Q. Who selects the exact position for placing the gun? *A.* The platoon commander.

Q. What arrangements are made when there is the possibility of aerial attack while on the march? *A.*

(1) Except for minor variations the drill and fire action on the march are the same as under other conditions. Time is not available to perform the duties prescribed under EXAMINE GUN.

(2) On the march the members of the gun section are stationed in the truck in positions to permit all around observation for aerial targets.

(3) Gun commanders are given instructions concerning the engagement of targets before the march begins. Fire is opened on hostile aircraft as soon as possible after they are observed except when fire is limited by previous instructions or the safety of adjacent troops. Whenever the gun commander hears the command or signal TARGET the truck is stopped and the gun prepared for action with maximum speed.

(4) As time is usually not available to set up central tracer control, firing during marches will normally be conducted by individual tracer control.

TABLE V.—*Drill for 37-mm gun section*

Details	DETAILS, POSTS	(a) EXAMINE GUN (b) REPORT	(a) TARGET (b) COMMENCE FIRING	(a) SUSPEND FIRING (b) CEASE FIRING
Gunner	Obtains vertical sight and places it in vertical sight seat. Takes post on right seat of carriage.	(a) Tests smoothness of operation of elevating handwheels. Turns leveling jack handle on left of carriage until level bubbles are centered. Assists gun commander in bore sighting gun and adjusting sighting system, if necessary; assisted by No. 1, then hooks up flexible cables and helps check synchronization of fire-control system. (b) Reports, "Sight and elevating mechanism in order."	(a) Trains his sight on designated target, calls out, "On target," and continuously tracks the target without further command. (b) Continues to track the target, and pushes down on trigger pedal with his foot. Keeps trigger pedal depressed until SUSPEND FIRING or CEASE FIRING is given.	(a) Lifts his foot from trigger pedal but continues to track the target. (b) Lifts his foot from the trigger pedal and ceases tracking, but remains at his post.
No. 1	Obtains lateral sight and places it in lateral sight seat. Then takes post on left seat of carriage.	(a) Tests smoothness of operation of traversing handwheels. Turns leveling jack handle on right of carriage until level bubble is centered. Assists gun commander in bore sighting and adjusting sighting system, if necessary; assists gunner in hooking up flexible cables and in checking synchronization of fire-control system. (b) Reports, "Sight and traversing mechanism in order."	(a) Trains his sight on designated target, calls out, "On target," and continuously tracks target without further command. (b) Continues to track the target.	(a) Continues to track the target. (b) Ceases tracking but remains at his post.
No. 2	Obtains the priming rod, a wrench, recoil oil, and waste and places them convenient to the gun. Takes post on right side of gun, opposite the breech and facing to the rear.	(a) Fills recuperator cylinder under immediate supervision of the gun commander. Examines chamber and bore, and if necessary, sponges them assisted by No. 5. (b) Reports, "Recoil mechanism and bore in order."	(a) When No. 3 calls "Prime," pulls lock frame to the rear and locked position with priming rod. (b) Receives empty ammunition clips from right side of breech and places them in a container clear of the gun. Keeps base of carriage free of empty cartridge cases. If a stoppage occurs, assists No. 3 in repeating the loading operation.	(a) Remains at his post. (b) Assists No. 3 in unloading the gun. When so directed by the gun commander he assists No. 3 in attaching water hose and cooling gun tube (barrel).

29

TABLE V.—*Drill for 37-mm gun section*—Continued

Details	DETAILS, POSTS	(a) EXAMINE GUN (b) REPORT	(a) TARGET (b) COMMENCE FIRING	(a) SUSPEND FIRING (b) CEASE FIRING
No. 3	Obtains lubricating oil and waste and places them convenient to the gun. Unlocks tube (barrel), and lowers gun support. Takes post on left side of gun, opposite the breech and facing to the rear at the ammunition tray.	(a) Examines water chest hose and if water is needed calls on Nos. 4 and 5 for assistance. Examines, cleans, and oils breech mechanism, breechblock, feed mechanism, and trigger mechanism. (b) Reports, "Gun and water chest in order."	(a) Opens feed box cover, calls "Prime," and holds carrier up until caught by carrier catch. When lock frame is pulled to the rear and locked position, calls, "Ammunition," receives a loaded ammunition clip from No. 4, places it, with cartridges down, on ammunition tray, and pushes it into the receiver against carrier catch. He then closes the feed box cover. (b) At the command LOAD given by the gun commander, pushes ammunition clip on info receiver until carrier is tripped, allowing lock frame to load the first round. Continues feeding loaded ammunition clips into receiver.	(a) Keeps one loaded ammunition clip in place on the ammunition tray. (b) Assisted by No. 2 unloads the gun, and when so directed by the gun commander attaches water hose and cools gun tube (barrel).
No. 4 (see note 1).	Assisted by No. 5, obtains designated number of loaded ammunition clips and places them near the ends of the two outriggers. Takes post on left side of gun and convenient to the ammunition tray.	(a) Assists Nos. 3 and 5 in obtaining water for the water chest when necessary. Assisted by Nos. 5 and 6 examines ammunition and loads additional ammunition clips if necessary. (b) Reports, "Ammunition in order."	(a) When No. 3 calls "Ammunition," hands him a loaded ammunition clip. Prepares to pass additional ammunition to No. 3. (b) Continues to pass ammunition to No. 3.	(a) Prepares to pass additional ammunition to No. 3. (b) Receives unfired ammunition from No. 3 and passes it to No. 5. Assisted by Nos. 5 and 6, replenishes supply of loaded ammunition clips, obtaining additional ammunition from the platoon when necessary.
No. 5 (see note 1).	Assists No. 4 in obtaining designated number of loaded ammunition clips and placing them near the ends of the two outriggers. Uncovers water chest. Takes post near the ammunition and convenient to No. 4.	(a) Assists Nos. 3 and 4 in obtaining water for the water chest when necessary. Assists No. 2 in sponging chamber and bore if necessary. Assists Nos. 4 and 6 in examining ammunition and loading additional ammunition clips if necessary. (b) No duties.	(a) Hands a loaded ammunition clip to No. 4 and prepares to pass additional clips as required. (b) Continues to pass ammunition to No. 4.	(a) Prepares to pass additional ammunition to No. 4. (b) Receives unfired ammunition from No. 4 and places it with the other clips of loaded ammunition. Assists Nos. 4 and 6 in replenishing supply of loaded ammunition clips and obtaining additional ammunition from the platoon when necessary.

| No. 6 (see note 1). | Carries designated number of boxes of ammunition to vicinity of gun position. Takes post at the ammunition cases, ready to load additional clips. | (a) Assists Nos. 4 and 5 in loading additional ammunition clips if necessary.
(b) No duties. | (a) No duties.
(b) Obtains empty ammunition clips from container near No. 2 and reloads them as rapidly as possible. | (a) Continues to reload empty ammunition clips.
(b) Assisted by Nos. 4 and 5, reloads empty ammunition clips. |

NOTES

1. In semimobile units, duties of No. 6 are performed by Nos. 4 and 5.
2. SUSPEND FIRING is employed when a brief halt in firing is desired. CEASE FIRING is employed when an appreciable period of time is to elapse before firing is resumed.

TABLE VI.—*Drill for emplacing 37-mm gun and march order*

Details	EMPLACING GUN	MARCH ORDER
Gunner	Assisted by No. 1, uncouples carriage from truck and sees that wheels are parallel to carriage side frames. (See note 2.) Unfastens the three gun cover straps on left of carriage and then assists in removing gun cover. Unfastens and removes one cover support. Grasps tripping and locking levers at front end of carriage. At the command TRIP, operates locking and tripping levers. Pushes down on carriage frame until it is on ground. Assists the men at counterpoise levers if they have difficulty in forcing their levers down or raising the wheels. (See note 3.) Assists in unloading matériel from truck.	Elevates or depresses gun until tube (barrel) can be engaged by gun support. When sighting system has been set at normal, disconnects flexible cables, assisted by No. 1. Removes vertical sight from sight seat and replaces it in chest provided. At the command READY, grasps tripping lever at front end of carriage. After carriage is raised, pushes tripping lever to the rear until locking lever is engaged. (See note 6.) Assists the men at counterpoise levers, if they have difficulty in lowering the wheels, raising the levers, or raising the carriage. (See note 5.) Replaces one cover support. Assists in replacing cover. Fastens the three gun cover straps on left of carriage. Assists in loading matériel on truck. When loading is completed, couples carriage to truck with assistance of No. 1.
No. 1	Assists gunner in uncoupling carriage from truck and in setting wheels parallel to carriage side frames. (See note 2.) Unfastens the three gun cover straps on right of carriage and then assists in removing gun cover. Unfastens and removes one cover support. Grasps tripping and locking levers at rear end of carriage. At the command TRIP, operates locking and tripping levers. Pushes down on carriage frame until it is on ground. Assists the men at counterpoise levers if they have difficulty in forcing their levers down or raising the wheels. (See note 3.) Assists in unloading matériel from truck.	Traverses gun until tube (barrel) can be engaged by gun support. When sighting system has been set, assists gunner in disconnecting flexible cables. Removes lateral sight from sight seat and replaces it in chest provided. At the command READY, grasps tripping levers at rear end of carriage. After carriage is raised, pulls tripping lever to the rear until locking lever is engaged. (See note 6.) Assists the men at counterpoise levers if they have difficulty in lowering the wheels, raising the levers, or raising the carriage. (See note 5.) Replaces one cover support. Assists in replacing cover. Fastens the three gun cover straps on right of carriage. Assists in loading matériel on truck. When loading is completed, assists gunner in coupling carriage to truck.
No. 2	Removes muzzle cover. When carriage is tripped, pushes down on hand-hold at right front of carriage until carriage is on ground. Takes post at right front wheel of carriage, extends counterpoise lever, and grasps lever with both hands. At the command LEVERS DOWN, forces lever down until wheel is in raised position. (See note 3.) Lowers and adjusts right outrigger. Assists in unloading matériel from truck.	Replaces and secures muzzle cover. Raises and locks right outrigger. Takes post at right front wheel of carriage. At the command READY, pushes down adjacent wheel until it makes contact with the ground and then grasps counterpoise lever with both hands. (See notes 4 and 5.) At the command LEVERS UP, lifts on the lever, rotating the cylinder about its center and forcing the rod down. At the command CARRIAGE UP, lifts on lever until carriage is raised and locked in traveling position. Assists in loading matériel on truck.
No. 3	Removes water chest cover. Unlocks tube (barrel) and lowers gun support sufficiently to allow gun cover to be removed. Then raises support and relocks tube. When carriage is tripped, pushes down on handhold at left rear of carriage until carriage is on ground. Takes post at left rear wheel of carriage, extends counterpoise lever, and grasps lever with both hands. At the command LEVERS DOWN, forces lever down until wheel is in raised position. (See note 3.) Lowers and adjusts left outrigger. Assists in unloading matériel from truck.	Replaces and secures water chest cover. Raises and locks left outrigger. Takes post at left rear wheel of carriage. At the command READY, pushes down adjacent wheel until it makes contact with the ground and then grasps counterpoise lever with both hands. (See notes 4 and 5.) At the command LEVERS UP, lifts on the lever, rotating the cylinder about its center, and forcing the rod down. At the command CARRIAGE UP, lifts on lever until carriage is raised and locked in traveling position. Unlocks the tube (barrel) and lowers gun support sufficiently to allow gun cover to be replaced. Then raises support and relocks tube. Assists in loading matériel on truck.

TABLE VI.—*Drill for emplacing 37-mm gun and march order*—Continued

Details	EMPLACING GUN	MARCH ORDER
No. 4	Releases gun cover lashings on left of carriage. Assists in removing gun cover. When carriage is tripped, pushes down on handhold at left front of carriage until carriage is on ground. Takes post at left front wheel of carriage, extends counterpoise lever, and grasps lever with both hands. At the command LEVERS DOWN, forces lever down until wheel is in raised position. (See note 3.) Assisted by No. 5, unloads ammunition from truck.	Raises gun support and locks the gun in its traveling position. Takes post at left front wheel of carriage. At the command READY, pushes down adjacent wheel until it makes contact with the ground and then grasps counterpoise lever with both hands. (See notes 4 and 5.) At the command LEVERS UP, lifts on the lever, rotating the cylinder about its center and forcing the rod down. At the command CARRIAGE UP, lifts on lever until carriage is raised and locked in traveling position. Assists in replacing gun cover. Fastens gun cover lashings on left of carriage. Assisted by No. 5, loads ammunition on truck.
No. 5	Releases gun cover lashings on right of carriage. Assists in removing gun cover. When carriage is tripped, pushes down on handhold at right rear of carriage until carriage is on ground. Takes post at right rear wheel of carriage, extends counterpoise lever, and grasps lever with both hands. At the command LEVERS DOWN, forces lever down until wheel is in raised position. (See note 3.) Assists No. 4 in unloading ammunition from truck.	Returns ammunition to ammunition chests. Takes post at right rear wheel of carriage. At the command READY, pushes down adjacent wheel until it makes contact with the ground and then grasps counterpoise lever with both hands. (See notes 4 and 5.) At the command LEVERS UP, lifts on the lever, rotating the cylinder about its center and forcing the rod down. At the command CARRIAGE UP, lifts on lever until carriage is raised and locked in traveling position. Assists in replacing gun cover. Fastens gun cover lashings on right of carriage. Assists No. 4 in loading ammunition on truck.
No. 6 (see note 1).	Assists where necessary in unloading truck. When all ammunition and matériel have been removed from truck, moves truck to designated parking area and then returns to the gun position.	Proceeds to truck park and drives truck to gun position. Lowers tailgate and assists in loading ammunition in truck.

NOTES

1. In semimobile units No. 6 is eliminated.

2. Setting the wheels parallel to the side frames of the carriage brings the lug on the steering arm tie rod in such a position that as the carriage drops into firing position the lug follows the slot on the buffer tie rod. While the carriage is down in firing position, the locking lever should engage the other end of the arc on the tripping lever. This will prevent the lug slipping out of the slot. However, before attempting to raise the carriage, a check should be made to see that the lug will follow the slot as the carriage comes up. If the lug does not follow the slot, the carriage will jam badly.

3. When lowering the carriage, standing or jumping on the counterpoise cylinder levers to increase the applied force is strictly prohibited since the increased force exerted may bend the counterpoise rods.

4. If the carriage has sunk into the ground the full length of the spade, difficulty may be encountered in pushing the wheels down far enough to allow the counterpoise cylinders to be rotated. Remove sufficient earth from under the wheels to allow them to be sufficiently depressed.

5. If the carriage has been emplaced on uneven ground, one or more of the wheels may snap up into their raised position as soon as the cannoneers release them to grasp the corresponding counterpoise levers. Where this occurs, the gun commander, the gunner, and No. 1 assist at the wheels where necessary by forcing the counterpoise rods down at the same time that the assigned cannoneers force the wheels down to the ground, and hold the counterpoise levers in position until the assigned cannoneers are able to grasp them.

6. The locking levers must be fully engaged whenever the carriage is in traveling position. Failure of the locking lever pin to fully engage in its seat may result in that end of the carriage concerned dropping to firing position while being towed. This may cause a serious accident. If this fault develops, either the locking pin seat should be reamed out, or the pin worked down with fine emery cloth until engagement is easy and positive.

6. **Drill of machine-gun squad.**—*Q*. What is the personnel of the machine-gun squad? *A*. A squad leader, 4 machine gunners, and 1 chauffeur.

Q. What is the formation for the machine-gun squad? *A*. See figure 10.

FRONT

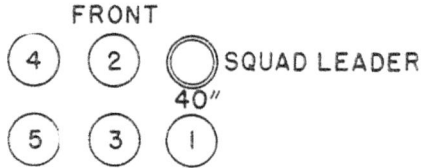

FIGURE 10.—Formation of machine-gun squad.

Q. What are the posts of the machine-gun squad? *A*. See figure 11.

Q. What are the duties of the members of the machine-gun squad at the various commands? *A*. See table VII.

Q. How is the gun loaded? *A*.

FIGURE 11.—Posts of machine-gun squad.

(1) No. 2 opens the ammunition chest and feeds the end of a belt into the receiver until one round is beyond the belt holding pawl. No. 3 raises the cover, lowering it as soon as No. 2 has fed the belt into the receiver. No. 1 pulls the bolt handle to the rear twice, releasing it each time from its rearmost position so as to allow the

driving spring to function. The gun is now loaded and ready to fire.

(2) In the case of the caliber .30 machine gun the following is the procedure: At the command LOAD, No. 1 advances his right hand opposite the belt exit and holds it in position ready to grasp the brass tag of the belt. No. 2 opens the ammunition chest, holds the end of the belt on the brass tag at the point where it joins the fabric and pushes the tag through the feed opening as far as possible. No. 1 grasps the tag as it is pushed from the belt exit and gives it a quick jerk to the right. He next pulls the bolt handle well to the rear and releases it. He again pulls the bolt handle to the rear and releases it. The gun is now loaded for automatic fire. No. 1 resumes his proper hold on the gun. The bolt handle should be released in its rearmost position so as to give the driving spring full play, and should not be eased forward.

Q. How is the gun unloaded? *A.* At the command UNLOAD, No. 1 unlatches and raises the cover with the left hand. No. 1 pulls back the bolt about 1 inch when No. 2 has raised the extractor. After the belt has been withdrawn, No. 1 pulls the bolt all the way back, releases it, pulls it back again, inspects the chamber to see that the gun is unloaded, releases the bolt handle, and pulls the trigger. As soon as No. 1 has raised the cover sufficiently, No. 2 raises the extractor, withdraws the belt, packs it carefully in the ammunition chest, and closes the ammunition chest cover.

Q. Name two instructions that must always be given at every drill. *A.* Whether or not live ammunition is to be put in the belts and whether or not the water jackets are to be filled.

Q. How is the piece inspected? *A.* The piece being in firing position, and the members of the machine-gun squad at their posts, as the inspector approaches, the squad leader commands: ATTENTION. faces the inspector, salutes, and reports, "Sir, Corporal _____, Battery ____, Squad leader." He then gives the necessary commands to demonstrate the mechanical functioning of the gun and mount. He gives such other commands as are necessary to execute the wishes of the inspector.

Q. Who selects the exact position for the guns? *A.* The platoon commander.

Q. What command is given by the squad leader to put the gun into firing position? *A.* PREPARE FOR ACTION.

Q. What command is given by the squad leader to check the gun, mount, and other equipment? *A.* EXAMINE GUN.

Q. What commands are given to conduct the fire? *A*. (1) LOAD, (2) TARGET, (3) COMMENCE FIRING, (4) CEASE FIRING (SUSPEND FIRING).

Q. What provisions are made when there is possibility of aerial attack while on the march? *A*.

(1) When there is any possibility of aerial attack during the movement, the gun will be mounted in the truck ready for prompt delivery of fire. The tripod or pedestal is set up and secured in the truck in such a manner that while it is held rigidly enough for traveling and firing, it may be dismounted from the truck without delay.

(2) Drill and fire action are conducted in the same manner and by the same commands as when in position off the truck, due allowance being made for the fact that the emplacement, the reserve ammunition, and the belt-filling station are all located together on the truck, and with the following exceptions:

(*a*) At the command or signal POSTS, the squad leader will at once command: LOAD. Nos. 1 and 2 load the gun and take positions which will enable them to maintain a sharp lookout and to open fire in the minimum time after the discovery of a hostile plane. If directed by the chief of section, the squad leader also requires other members of the squad to keep a careful watch in the directions assigned to them (rear, sides, front). Any member of the squad sighting a hostile plane will at once call out TARGET, and point to the plane. The squad leader will signal TARGET, by whistle.

(*b*) Gunners are usually given their instructions relative to the engagement of the target prior to the beginning of a movement. Limited only by such instructions, gunners open fire on hostile targets at the earliest moment without command, and cease fire only on orders or when the target passes out of the field. Squad leaders, in addition to their usual functions during firing, are responsible that only suitable targets are engaged and that advance instructions as to fire action are carried out. They suspend or cease the firing when necessary and execute any commands received. In cases where it is practicable for the platoon commander to direct the fire of his platoon the procedure is as nearly as possible the same as when in position.

(*c*) During action squad leaders dispose their men in such a manner as to interfere as little as possible with fire, to get as much shelter as possible, and to facilitate the supply of ammunition and water to No. 2 as he calls for it.

Table VII.—Drill for machine-gun squad (caliber .50 machine gun)[1]

Details	PREPARE FOR ACTION[2]	EXAMINE GUNS[3]	(a) LOAD (b) TARGET[4]	COMMENCE FIRING	(a) SUSPEND FIRING[5] (b) CEASE FIRING[5]	MARCH ORDER[2]
No. 1	Drill A.—Assisted by No. 3, he removes the ammunition chest from the mount and the link bag from the link chute, dismounts the gun, passes the gun off the rear of the truck to No. 2, removes the cradle, and carries the cradle to the gun position, following No. 2. When the pedestal and legs are assembled, he mounts the cradle, assisted by No. 3. Then, assisted by No. 2, he mounts the gun, boresights the gun, if necessary, and aids the platoon headquarters and range section in attaching the central control cables, if used. He takes post in rear of the gun. Drill B.—Assisted by No. 3, he removes the ammunition chest from the mount, and the link bag from the link chute, dismounts the gun, and passes the gun off the rear of the truck to No. 2. When the yoke has been removed, Nos. 4 and 5 assist No. 1 in folding up the tripod, removing it from the truck, and carrying it, behind No. 3, to the position. Upon arrival at	He sees that the gun mechanism and the mount are in good working order and adjustment, that the sighting mechanism, if used, operates properly, and that the gun elevates and traverses freely. He then reports, "Gun and mount in order."	(a) At the command LOAD, as soon as the ammunition belt is fed into the receiver, he pulls the bolt to the rear twice. (b) At the command TARGET, he begins to track the target by traversing and elevating or depressing the gun and then calls out, "On target."	He continuously presses the trigger, and tracks the target with his sights (central control) or with the aid of trackers (individual tracker control). If stoppages occur, he clears them with the aid of Nos. 2 and 3.	(a) At the command SUSPEND FIRING, he releases the trigger and continues to track the target. (b) At the command CEASE FIRING, he releases the trigger and stops tracking the target. With the aid of No. 2, he unloads the gun.	Drill A.—Assists the platoon headquarters and range section in detaching the central control cables, if used. Assisted by No. 3, removes the ammunition chest from the mount, and the link bag from the link chute, dismounts the gun, passes the gun to No. 2, removes the cradle, and carries the cradle to the truck, following No. 4. When the pedestal and legs are assembled, he mounts the cradle, assisted by No. 3. He then mounts the gun, assisted by No. 2. He takes post in the truck convenient to the gun. Drill B.—Assisted by No. 3, removes the ammunition chest from the mount and the link bag from the link chute, dismounts the gun, and passes the gun to No. 2. When the yoke has been removed, he is assisted by Nos. 4 and 5 in folding up the tripod, removing it to the truck, followed by the squad leader, and setting it up in and securing it to the truck. He then mounts the gun assisted by No. 2. He then takes post in the truck convenient to the gun.

TABLE VII.—*Drill for machine-gun squad (caliber .50 machine gun)*—Continued

Details	PREPARE FOR ACTION	EXAMINE GUNS	(a) LOAD (b) TARGET	COMMENCE FIRING	(a) SUSPEND FIRING (b) CEASE FIRING	MARCH ORDER
No. 1.—Con.	the position, No. 1, assisted by Nos. 4 and 5, sets up the tripod. He then mounts the gun with the assistance of No. 2. He takes post in rear of the gun.					
No. 2	Drill A.—He jumps off the truck, receives the gun from Nos. 1 and 3, and proceeds to the gun position. When the cradle is mounted, he assists No. 1 in mounting the gun and bore-sighting the gun, if necessary, and aids in the attachment of the central control cables, if used. He then assists Nos. 3 and 4 in bringing up the remaining equipment. He takes post on the left of the gun and facing it. Drill B.—He jumps off the truck, receives the gun from Nos. 1 and 3, and proceeds to the position. When the yoke is mounted, he assists No. 1 in mounting the gun. He then assists Nos. 3 and 4 in bringing up the remaining equipment. He takes post on the left of the gun and facing it.	He sees that there is one ammunition chest on the gun and that a full ammunition chest is located convenient to the gun. When no firing is to occur, he sees that there is no live ammunition near the gun. He then reports, "Ammunition in order."	(a) At the command LOAD, he opens the ammunition chest and feeds the end of the ammunition belt into the receiver until one round is beyond the belt-holding pawl. (b) At the command TARGET, he shifts his post to conform to the movement of the gun.	If stoppages occur, he assists Nos. 1 and 3 in clearing them. Assisted by No. 5, he replaces the ammunition chest on the gun with additional chests as required.	(a) Same as under TARGET. (b) When the cradle is clamped, he assists No. 1 in unloading the gun. He replaces the belt in the ammunition chest and closes the top of the chest. He replaces the ammunition chest, if necessary.	Drill A.—Assists the platoon headquarters and range section in detaching the central control cables, if used. He receives the gun from Nos. 1 and 3. When the mount is disassembled, he follows Nos. 1 and 3 to the truck. When the mount is assembled in the truck he assists No. 1 in mounting the gun. He then assists Nos. 3 and 4 in bringing up the remaining equipment from the position to the truck. He takes post on the truck as directed by the squad leader. Drill B.—This drill is the same as drill A except that he follows No. 3 in carrying the gun to the truck.

No. 3

Drill A.—He assists No. 1 in removing the ammunition chest from the mount and the link bag from the link chute, dismounting the gun, passing the gun to No. 2, dismounting the cradle, carrying the cradle to the position, and mounting the cradle. He then returns to the truck for an ammunition chest and the link bag, carries them to the position, and attaches the remaining equipment to the position. He then takes post on the right of the gun and facing it.

Drill B.—He assists No. 1 in removing the ammunition chest from the mount, and the link bag from the link chute, dismounting the gun, and passing the gun to No. 2. He dismounts the yoke and follows No. 2 to the position. When the tripod is set up, he mounts the yoke. He then returns to the truck for an ammunition chest, and the link bag, carries it to the position, and attaches it to the mount. He assists Nos. 2 and 4 in carrying the remaining equipment to the position. He then takes post on the right of the gun and facing it.

He checks the cables, if used, to see that there are no kinks and that there is sufficient slack. He then reports, "Cables in order."

(a) At the command LOAD, he raises the cover and then unclamps the cradle. When No. 2 has fed the ammunition belt into the receiver, he lowers the cover. When loading the caliber .30 gun, he assists in loading by pulling on the brass tab of the belt until the first round is engaged.

(b) At the command TARGET, he shifts his post to conform to the movement of the gun.

If stoppages occur, he assists Nos. 1 and 2 in clearing them. He removes, empties, and replaces the link bag when necessary.

(a) Same as under TARGET.

(b) When No. 1 stops tracking, No. 3 clamps the cradle. He removes the link bag, empties the links into a suitable receptacle, and then replaces the link bag. He collects the empty cartridges and places them in suitable receptacles.

Drill A.—He assists No. 1 in removing the ammunition chest from the mount, and the link bag from the link chute, dismounting the gun, passing the gun to No. 2, dismounting the cradle, carrying the cradle to the truck, and mounting the cradle. He then returns to the position for an ammunition chest and the link bag, carries them to the truck, and attaches them to the mount. He then assists Nos. 2 and 4 in carrying the remainder of the equipment to the truck. He takes post on the truck as directed by the squad leader.

Drill B.—He assists No. 1 in removing the ammunition chest from the mount, and the link bag from the link chute, dismounting the gun, and passing the gun to No. 2. He dismounts the yoke, follows Nos. 1, 4, and 5 to the truck, and when the tripod is set up, mounts the yoke. He then returns to the position for an ammunition chest and the link bag, carries them to the truck, and attaches them to the mount. He then assists Nos. 2 and 4 in carrying the remainder of the equipment to the truck. He takes post on the truck as directed by the squad leader.

Table VII.—*Drill for machine-gun squad (caliber .50 machine gun)*—Continued

Details	PREPARE FOR ACTION	EXAMINE GUNS	(a) LOAD (b) TARGET	COMMENCE FIRING	(a) SUSPEND FIRING (b) CEASE FIRING	MARCH ORDER
No. 4	*Drill A.*—He disconnects the water hoses and coils them on the water chest. Assisted by No. 5, he removes the legs from the pedestal. He takes the pedestal off the rear of the truck and follows Nos. 1 and 3 to the gun position. Assisted by No. 5, he assembles the pedestal and legs. He returns to the truck for the water chest and hoses, carries them to the position, assisted by No. 5, and connects the water hoses to the gun. He then assists Nos. 2 and 3 in carrying the remaining equipment to the position. He takes post at the water pump. *Drill B.*—He disconnects the water hoses and coils them on the water chest. He assists Nos. 1 and 5 in folding up the tripod, removing it from the truck, carrying it to the position, and setting it up. He returns to the truck for the water chest and hoses and, assisted by No. 5, carries them to the position. He then connects the water hoses to the gun. He assists Nos. 2 and 3 in carrying the remaining equipment to the posi-	He sees that the water jacket is filled, that all water pipes and hose connections are tight, and that the water chest is filled and in working order. He also sees that a full water container is conveniently located. He then reports: "Water pump in order."	(a) He remains at the water pump. (b) If necessary, he shifts the position of the water chest to avoid being in front of the muzzle when fire is commenced.	He continuously operates the water pump.	(a) Continues to operate the water pump. (b) Continues to operate the water pump for about 20 full turns of the pumphandle after the gun ceases firing. If necessary, he changes or replenishes the water in the water chest.	*Drill A.*—He disconnects the water hoses and coils them on the water chest. Assisted by No. 5, he removes the legs from the pedestal. He then carries the pedestal to the truck, following No. 5. With the aid of No. 5, he assembles the pedestal and legs, and sets the mount up in and secures it to the truck. He returns to the position for the water chest and hoses, carries them assisted by No. 5 to the truck, and connects the water hoses to the gun. He then returns to the position to assist Nos. 2 and 3 in carrying the remainder of the equipment to the truck. He takes post on the truck convenient to the water pump. *Drill B.*—He disconnects the water hoses and coils them on the water chest. He assists Nos. 1 and 5 in folding up the tripod, removing it to the truck, and setting it up in and securing it to the truck. He returns to the position for the water chest and hoses, carries them to the truck, and connects the water hoses to the gun. He then returns to the

tion. He then takes post at the water pump.

position to assist Nos. 2 and 3 in carrying the remainder of the equipment to the truck. He takes post on the truck convenient to the water pump.

Drill A.—He obtains his truck, maneuvers it as close to the gun position as circumstances permit, and lowers the tailgate. He then returns to the position and assists No. 4 in removing the legs from the pedestal. He carries the legs to the truck, following the squad leader. He assists No. 4 in assembling the pedestal and legs and setting the mount up in and securing it to the truck. He assists No. 4 to carry the water chest and hoses to the truck. When all men and equipment are in the truck, he closes the tailgate. He then maneuvers his truck as directed.

Drill B.—He obtains his truck, maneuvers it as close to the gun position as circumstances permit, and lowers the tailgate. He assists Nos. 1 and 4 in folding up the tripod, carrying it to the truck, and setting it up in and securing it to the truck. He assists No. 1 to carry the water chest and hoses to the truck. When all men and equipment are in the truck, he closes the tailgate. He then maneuvers his truck as directed.

No. 5	Drill A.—He maneuvers his truck as close to the designated position as circumstances permit, dismounts, and lowers the tailgate. He assists No. 4 in removing the legs from the pedestal. He carries the legs to the position, following No. 4. At the position he assists No. 4 in assembling the pedestal and legs. He then returns to the truck and assists No. 4 in carrying the water chest and hoses to the gun position. When all equipment is removed from the truck, he puts up the tailgate, moves the truck to the designated parking area, and then returns to the gun position. He takes post near the reserve ammunition. Drill B.—He maneuvers his truck as close to the designated position as circumstances permit, dismounts, and lowers the tailgate. He assists Nos. 1 and 4 in folding up the tripod, removing it from the truck, carrying it to the position, and setting it up. He then returns to the truck and assists No. 4 in carrying the water chest and hoses to the position. When all equip-	He inspects the reserve ammunition and the link-loading machine, and if no firing is to occur, sees that no live ammunition is near the reserve chests. He then reports, "Reserve ammunition in order."	(a) Remains at the reserve ammunition chests. (b) Remains at the reserve ammunition chests.	He brings up reserve ammunition chests to the gun when called for by No. 2 and assists him in removing the old ammunition chest and mounting the new one. He loads additional belts, and obtains ammunition from the platoon when required.	(a) Continues as under COMMENCE FIRING. (b) Replaces the empty ammunition chest with a full one, if necessary. He loads belts with the assistance of other members of the squad until all ammunition chests are full.

41

TABLE VII.—*Drill for machine-gun squad (caliber .50 machine gun)*—Continued

Details	PREPARE FOR ACTION	EXAMINE GUNS	(a) LOAD (b) TARGET	COMMENCE FIRING	(a) SUSPEND FIRING (b) CEASE FIRING	MARCH ORDER
	ment is removed from the truck, he puts up the tailgate, moves the truck to the designated parking area, and then returns to the gun position. He takes post near the reserve ammunition.					

NOTES

[1] The drill, with minor changes caused by differences in matériel, is also suitable for units equipped with caliber .30 machine guns.

[2] Drill A is for the M2 antiaircraft machine-gun mount. Drill B is for the M1 antiaircraft machine-gun mount.

[3] When the truck park is a considerable distance from the emplacement, the squad proceeds with EXAMINE GUNS without waiting for No. 5, his duties being performed by No. 3.

[4] In assigning the target in units which normally employ central tracer control, the command, TARGET—GUNNER'S ACTION, is given if it is desired to employ individual tracer control.

[5] SUSPEND FIRING is employed when a brief halt in firing is desired. CEASE FIRING is employed when an appreciable period of time is to elapse before firing is resumed.

[6] No. 6 in the second squad of each section loads ammunition into belts continuously during and between actions, when necessary. At other times, he performs such duties as may be prescribed by the chief of section.

CHAPTER 3

GUN AND MOUNT

SECTION I

105-MM GUN

7. Nomenclature.—*Q.* Point out and state the purpose of the principal parts of the gun as follows:

Breechblock.

Breech-operating handle.

Breech ring.

Closing chain.

Closing spring.

Extractors.

Firing mechanism.

Operating shaft.

Recoil lug.

Tube.

A. Practical demonstration on the gun. See also figures 12, 13, 14, and 15.

FIGURE 12.—105-mm gun (right side).

FIGURE 13.—105-mm gun (left side).

A. Left cradle extension.
B. Operating cam.
C. Rammer head.
D. Rammer body.

E. Cam (roller part for kicker).
F. Loading tray.
G. Rammer tripping cam.
H. Rammer latch release levers.

FIGURE 14.—105-mm gun loading rammer mechanism. (Gun removed; loading tray drawn to rear to expose cam E.)

A. Racer.
B. Pedestal.
C. Traversing mechanism.
D. Azimuth receiver drive mechanism.

E. Base plate.
F. Roller assembly.
G. Firing clips.

FIGURE 15.—105-mm gun mount.

Q. Point out and state the purpose of the principal parts of the mount as follows:

Azimuth indicator.
Base plate.
Breech-operating cam.
Breech-operating platform.
Counterrecoil cylinders.
Cradle.
Cradle extensions (right and left).
Elevating handwheels.
Elevation indicator.
Fuze range indicator.
Fuze setter.

Kicker.
Loading tray.
Pedestal.
Racer.
Rammer body.
Rammer buffers (front and rear).
Rammer head.
Rammer latch release lever.
Rammer retracting mechanism.
Rammer tripping cam.
Recoil cylinder.

Right and left yokes.   Traversing handwheels.

Roller assembly.   Working platform.

A. Practical demonstration on the gun. See also figures 12, 13, 14, and 15.

8. Action and minor adjustments.—*Q.* What is the purpose of the loading rammer mechanism? *A.* It is for the purpose of ramming the round into the gun.

Q. By what means is the rammer operated? *A.* By the action of compressed air.

Q. When the rammer is in the forward position, what should be the air pressure? *A.* The reading with the rammer forward and the gun cool (not recently fired), should be approximately 80 pounds per square inch.

Q. What trips the breechlock to close it? *A.* The rim of the cartridge case striking the extractors as the round enters the chamber.

Q. Explain the action of the rammer head at the end of the ramming stroke. *A.* The breechblock, in closing, strikes the rammer head and forces it up in its guide until it is latched in the upper position by its spring plunger.

Q. How is the breech opened? *A.* During the counterrecoil of the gun, the crank on the left end of the operating shaft strikes the lower surface of the operating cam. This forces the operating shaft to rotate and open the breech. The breech may also be opened by pulling the breech operating handle to the rear and downward.

Q. What happens to the rammer during recoil? *A.* The rammer is thrown to the rear by the action of the kicker until the rammer latch engages the rack on the forward end of the rammer, thus holding the rammer in the retracted or loading position.

Q. Can the rammer mechanism be tripped when the breechblock is closed? *A.* No. A safety device on the cradle prevents the release of the retracted rammer while the breechblock is closed. Opening of the breechblock cuts out the device and permits the release of the rammer.

Q. How is the rammer mechanism set to stop the ejected cartridge case in the loading tray? *A.* The rammer tripping cam is turned to its "engaged" position. When in this position, it is struck by a crank on the top of the rammer head during the last portion of the movement of the rammer body to the rear. This operation disengages the spring plunger which locked the rammer head in its upper position and permits the rammer head spring to push the rammer back to its lower position, in which position it stops the ejected cartridge case in the loading tray.

Q. Where is the recoil cylinder mounted? *A.* Underneath the center of the cradle.

Q. What takes up the energy of recoil? *A.* The oil in the recoil cylinder.

Q. Is the length of recoil fixed or variable? *A.* Fixed.

Q. After the gun has recoiled, what forces the gun to return to battery? *A.* The springs in the counterrecoil cylinders.

Q. How many such cylinders are there? *A.* Two.

Q. Where are they located? *A.* On the bottom of the cradle, one on each side of the recoil cylinder.

Q. What is the purpose of the counterrecoil buffer? *A.* To slow up the last portion of the movement of the gun in counterrecoil, thus permitting the gun to return to battery without shock.

Q. Where is the counterrecoil buffer located? *A.* In the forward end of the recoil cylinder.

Q. What kind of firing mechanism is used? *A.* The self-cocking, continuous-pull type. When the lanyard is released, the firing pin automatically returns to the cocked position.

Q. Where is the firing mechanism? *A.* In the breechblock.

Q. Is this mount provided with a traversing clutch so that the carriage may be traversed rapidly by hand? *A.* No.

Q. How many traversing speeds are provided? *A.* Two. The slow traversing motion gives 1.236° traverse for one turn of the handwheels, and the fast motion gives 3.75° traverse for one turn of the handwheels.

Q. How many elevating speeds are provided? *A.* One. One turn of the elevating handwheels gives an elevation (or depression) of 1.875°.

Q. How is the recoil cylinder filled? *A.* Elevate the gun to 80°. Remove the filling plug and vent plug and fill recoil cylinder to overflowing, using filling funnel provided. Replace filling plug and vent plug.

Q. How is the recoil cylinder drained? *A.* With gun horizontal, place a receptacle under the drain plug and remove the plug. If oil does not flow freely, remove the vent plug. After oil is drained, replace plugs.

Q. How is the oil seal in the floating piston of the rammer mechanism replenished? *A.* Remove the rammer piston valve body cap. Fill the oil screw filler with heavy recoil oil and attach it to the bayonet type fitting provided. Force the oil in by turning the handle of the filler until filler is emptied.

Q. How is the front buffer cylinder of the rammer filled? *A*. Depress the gun as far as possible. Remove the filling plug on top of the front buffer cylinder. Fill the buffer mechanism with heavy recoil oil and replace the plug. It is imperative that the buffer be kept full of oil at all times.

Q. How is the rear buffer cylinder of the rammer filled? *A*. Set the gun at 0° elevation. Remove the pipe plug and fill with heavy recoil oil until the oil is level with the filling plug hole. Replace the pipe plug.

Q. What kind of oil is used in the recoil and rammer mechanisms? *A*. Heavy recoil oil, low pour point (U. S. Army Spec. 2–96).

For data comparing the performance of the 105-mm gun with the 3-inch gun, see paragraph 10.

9. Care of gun and mount.—*Q*. What care should be given the bore of the gun after firing? *A*. As soon as possible after firing, the bore should be washed with a solution of ½ to 1 pound of sal soda

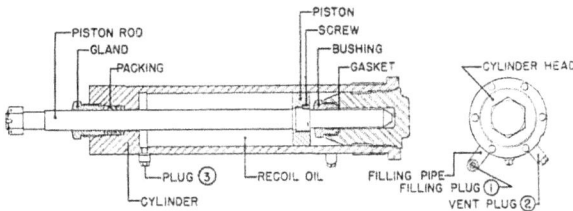

FIGURE 16.—Recoil cylinder.

(soda ash) per gallon of boiling water, a sponge being used for swabbing purposes. Special attention should be given to that portion of the bore extending from the origin to a point about 24 inches forward, as most of the fouling takes place in that area. Cleaning should be followed by thorough drying with the sponge covered with burlap, after which the bore should be oiled with a light coat of rust-preventive compound applied with the bore slush brush.

Q. What are some of the routine checks that should be made daily? *A*. (1) Open and close breech to see that it operates freely.

(2) Examine breech recess and bore to see that they are clean.

(3) See that firing mechanism works freely.

(4) Elevate and depress gun to see that the mechanism operates without binding or undue lost motion.

(5) Traverse gun to the right and left through the full extent of its travel to see that the mechanism operates without binding or undue lost motion.

(6) See that sliding surfaces of the gun and cradle are clean and well lubricated.

(7) See that all working parts are thoroughly lubricated.

(8) Examine recoil system for all oil leaks.

(9) Examine all keys, thongs, and hinges to see that they are in serviceable condition.

(10) Check tools and accessories to see that they are in their proper position and none is missing.

Q. Of what does maintenance of guns consist? *A.* Routine checks as stated in the answer to the previous question, lubrication at periodic intervals, and cleaning and painting.

Q. Where are cleaning and preserving materials obtained? *A.* They are issued to the battery by the regimental supply officer on the basis of an allowance prescribed in Ordnance Standard Nomenclature Lists or Army Regulations. Supplies are usually drawn quarterly.

Q. What care should be given to the breech mechanism after firing? *A.* The breech mechanism should be disassembled immediately after firing and cleaned and oiled. In case the mechanism is to be left unused for a considerable length of time, all bright surfaces should be coated with a rust-preventive compound.

Q. What care should be given to the firing mechanism after firing? *A.* The firing mechanism should be disassembled after firing. All burs and rough surfaces should be removed with a smooth file. The parts should be washed in dry cleaning solvent and wiped dry. When reassembling, the parts should be lightly coated with light class D lubricating oil. (Fed. Spec. VV-0-496.)

Q. What will be the result if oil thicker than authorized is used for coating the firing mechanism? *A.* The heavy oil will cause the energy of the firing spring to be absorbed, resulting in a misfire. This is especially true in cold weather when a thicker oil tends to congeal.

Q. What precautions should be used to prevent the entrance of foreign particles into the recoil system when filling with oil? *A.* The oil should be strained through a fine clean cloth, and the receptacles used in handling the oil should be clean.

Q. What care should be given the fuze setter M9? *A.* Before operation, the fuze setter should be opened and all parts accessible should be cleaned with clean, lint-free cloths. It should then be well lubricated. Lubricating facilities are painted red for identification. Light lubricating oil, class A, should be used in the oilholes provided and for all moving surfaces in contact, including hooks, levers, latches, and plungers. Interior parts not accessible should be flushed out with the lubricating oil through the numerous vents provided. The ball bearings in the hub of the indicator bracket are packed with a mixture of 60 percent petrolatum (U. S. Army Spec. 2-67), and 40 percent rust-preventive compound, grade "B" medium (U. S. Army Spec. 2-78A). They are packed during initial assembly and should not require further attention for several years. Repacking should be done by ordnance personnel.

SECTION II

3-INCH GUN

	Paragraph
General	10
Nomenclature, fixed mount	11
Action and minor adjustments, fixed mount	12
Nomenclature, mobile mount	13
Action and minor adjustments, mobile mount	14
Care of gun and mount	15

10. **General.**—*Q.* Compare the 3-inch gun (mobile) with the 3-inch gun (fixed) and the 105-mm gun (fixed). *A.* See following table:

Item	3-inch gun (mobile) M3	3-inch gun (fixed) M1917	105-mm gun M3
Weight of gun and mount, complete.	16,000 lbs	15,000 lbs	33,500 lbs.
Length of bore	Cal. .50	Cal. .55	Cal. .60.
Muzzle velocity, f/s (shrapnel)	2,600	2,600	No shrapnel used.
Muzzle velocity, f/s (HE shell)	2,700	2,700	2,800.
Limits of elevation	−1° to +80°	−5° to +85°	−5° to +80°.
Traverse	360°	360°	360°.
Maximum horizontal range (shrapnel).[1]	7,550 yds	7,550 yds	
Maximum horizontal range (HE shell).[2]	11,100 yds	11,100 yds	13,100 yds.
Maximum vertical range (shrapnel).[1]	8,600 yds	8,600 yds	
Maximum vertical range (HE shell).[2]	9,800 yds	9,800 yds	12,300 yds.
Type of recoil mechanism	Constant	Constant	Constant.
Type of counterrecoil mechanism.	Hydropneumatic.	Hydrospring.	Hydrospring.
Method of loading	Hand	Hand	Power rammer.

[1] Limited by power-train time fuze.   [2] Limited by mechanical time fuze.

Q. What trips the breechblock to close it? *A.* The rim of the cartridge case striking the extractors as the round enters the chamber.

Q. What causes the breech to open automatically? *A.* When the gun returns to battery during counterrecoil the operating crank of the breech strikes the operating cam on the carriage and opens the breech.

BREECH OPERATING MECHANISM

COUNTER-RECOIL CYLINDER

RECEPTACLE FOR FUZE SETTER CABLE

GUN JUNCTION BOX

LIGHT SWITCH

TRAVERSING HANDWHEEL

AZIMUTH INDICATOR

CRADLE

TRUNNION

TUBE

SPEED CHANGE LEVER

FIGURE 17.—3-inch gun, fixed mount (left side).

Q. Can the breech also be opened by hand? *A.* Yes, by means of the operating handle.

Q. What kind of firing mechanism is used? *A.* The self-cocking, continuous-pull type. When the lanyard is released, the firing pin automatically returns to the cocked position.

Q. If the gun does not fire at the first pull of the lanyard, does the firing mechanism have to be cocked before attempting to fire again? *A.* No, not as a separate action. It is merely necessary to pull the lanyard again.

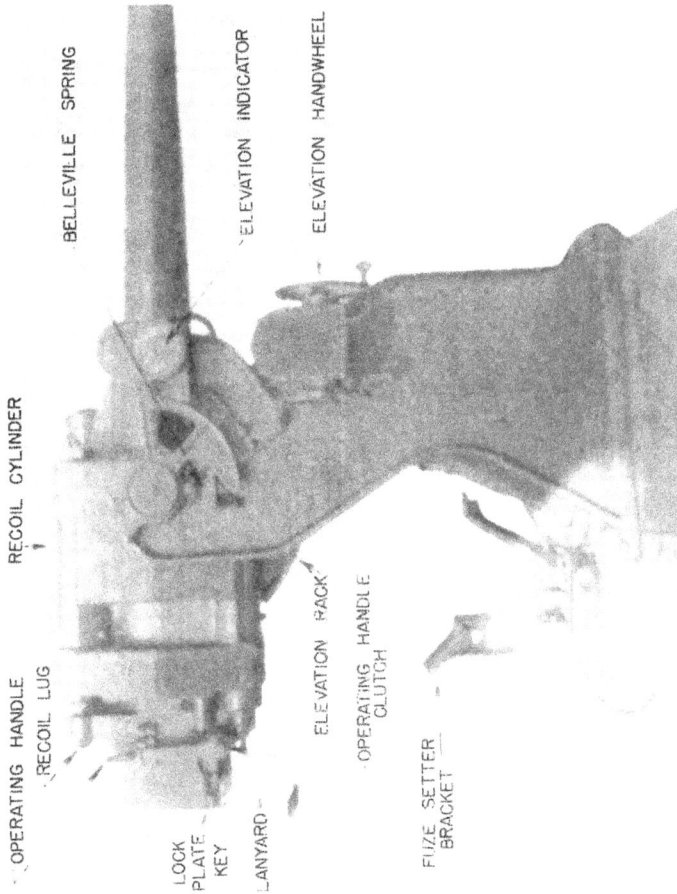

FIGURE 18.—3 inch gun, fixed mount (right side).

Q. Where is the firing mechanism located? *A.* In the breechblock.

Q. What ejects the empty cartridge case? *A.* The extractors throw it out automatically as the breechblock opens fully.

Q. How is the breechblock set for semi-automatic operation? *A.* The operating cam is set to the upper position so that it will engage

53

the lug on the end of the operating shaft as the gun moves in recoil and counterrecoil.

11. Nomenclature, fixed mount.

Q. Point out and state the purpose of the principal parts of the gun as follows:

Breechblock.	Liner retaining ring.
Breech recess.	Lock plate.
Chamber.	Lock plate key.
Closing chain.	Operating handle.
Closing spring.	Operating handle clutch.
Extractors.	Operating handle latch.
Firing mechanism.	Recoil lug.
Jacket.	Trigger shaft.
Lanyard.	Tube.
Liner.	

A. Practical demonstration on the gun. See also figures 17 and 18.

Q. Point out and state the use of the principal parts of the mount as follows:

Antifriction elevation device.	Fuze setter bracket.
Azimuth circle.	Fuze setter light.
Azimuth indicator.	Fuze setting crank.
Azimuth orienting knob.	Gun junction box.
Azimuth synchronizing screws.	Light switch.
Counterrecoil cylinders.	Mechanical pointers.
Cradle.	Oil filling hole.
Electrical pointers.	Oil filling valve.
Elevating gear.	Piston rod.
Elevation handwheel.	Racer.
Elevation indicator.	Recoil cylinder.
Elevation orienting knob.	Recuperator air filling valve.
Elevation synchronizing	Release lever (fuze setter).
screws.	Sliding covers (on indicators).
Fuze adjusting handwheel.	Traversing handwheel.
Fuze range indicator.	Trunnions.
Fuze setter.	

A. Practical demonstration on the gun. See also figures 17 and 18.

12. Action and minor adjustments, fixed mount.—*Q.* Where is the recoil cylinder located? *A.*

(1) On both the M1917M1 and the M3 mounts there are two recoil cylinders and two counterrecoil cylinders. They are mounted in pairs above and below the gun. The upper left and lower right ones are recoil cylinders.

(2) On the M1917 and the M1917MII mounts there is a single recoil cylinder mounted on top.

Q. How does the recoil cylinder check the recoil? *A*. The recoil cylinder is attached to the mount and contains a piston attached to the gun. The cylinder is filled with recoil oil. The recoil cylinder has throttling grooves cut in its walls to permit oil to pass around from one side of the piston to the other, as the piston is moved. The area of the orifice at any point, for passage of the oil, is controlled by varying the depth of the throttling grooves. By this method, the passage of the recoil oil is controlled so that the resistance to the movement of the piston increases as the recoil approaches its maximum, thereby gradually decreasing the speed of recoil.

Q. What care do the recoil cylinders require? *A*. Recoil cylinders should be emptied and refilled at least once every 3 months. They should be cleaned by ordnance personnel at least once every 6 months.

Q. How do you drain and refill the recoil cylinder? *A*. Remove filling and drain plugs and allow oil to drain out. With the gun in a horizontal position and the drain plugs replaced tightly, fill recoil cylinders with light recoil oil through the filling hole. Leave a void of not in excess of one-fourth of a pint of oil in the cylinder to allow for expansion of the oil as it becomes heated during firing.

Q. How do the counterrecoil cylinders work? *A*. When the gun recoils it compresses the springs in the counterrecoil cylinders. As soon as the gun has stopped in recoil the springs push against the counterrecoil pistons and force the gun back into battery.

Q. What are the cylinders fastened to? *A*. To the cradle. They do not move with the gun in recoil.

Q. What are the pistons fastened to? *A*. To the piston rods, which are fastened to the recoil lug and breech ring. They move with the gun as it recoils.

Q. What keeps the gun from returning to battery with a shock? *A*. The counterrecoil buffer.

Q. Where is the counterrecoil buffer located? *A*. In the forward end of the recoil cylinder.

Q. How many counterrecoil springs are there? *A*. Two in each cylinder, an inner and an outer.

Q. What type of top carriage has this gun? *A*. It is of the racer and traversing roller type. There is a pintle in the center and 30 rollers between the roller path on the base plate and the lower side of the racer, which bears on them.

Q. Can the mount be traversed slowly or rapidly at will? *A.* Yes, there are two-speed gears on the traversing shaft. They are shifted by a lever.

Q. What type of fuze setter does this gun use? *A.* The fuze setter M8.

FIGURE 19.—3-inch gun M3 on M2A2 mount (right side).

FIGURE 20.—3-inch gun M3 on M2A2 mount (left side).

Labels: OIL SCREW FILLER BRACKET — BREECH OPER-ATING MECH-ANISM — TRIP LEVER — SETTING CRANK — OIL FILLING VALVE — ADJUSTING HANDWHEEL — BRACKET FUZE SETTER — TRUNNION CRADLE — BREECH LIGHT — OIL FILLING HOLE — LIGHT SWITCH — GUN JUNCTION BOX — TROUBLE LAMP PLUG — RECOIL CYLINDER — TUBE — AZIMUTH DATA INDICATOR — EQUILIBRATOR AIR FILLING VALVE — TRAVERSING HANDWHEEL — GUN RECEPTACLE BOX — FUZE DATA INDICATOR — LEVELING JACKS

13. Nomenclature, mobile mount.

Q. Point out and state the purpose of the principal parts of the gun as follows:

Breechblock.
Breech recess.
Chamber.
Closing chain.
Closing spring.
Extractors.
Firing mechanism.
Jacket.
Lanyard.
Liner.

Liner retaining ring.
Lock plate.
Lock plate key.
Operating handle.
Operating handle clutch.
Operating handle latch.
Recoil lug.
Trigger shaft.
Tube.

A. Practical demonstration on gun. See also figures 19 and 20.

Q. Point out and state the use of the principal parts of the mount as follows:

Air reservoir cylinder.
Azimuth indicator.
Azimuth orienting knob.
A z i m u t h synchronizing screw.
Bogie clamping screws.
Breakaway switch (front bogie).
Breech recess light.
Buffer oil filling plug.
Counterrecoil buffer cylinder.
Counterrecoil cylinders.
Cradle.
Electrical pointers.
Elevating gear.
Elevating handwheel.
Elevation indicator.
Elevation orienting knob.
E l e v a t i o n synchronizing screw.
Equilibrator air filling valve.
Equilibrator cylinder.
Floating piston cylinder.
Fuze adjusting handwheel.
Fuze range indicator.
Fuze setter.

Fuze setter bracket.
Fuze setter light.
Fuze setting crank.
Gun junction box.
Jack support channel.
Leveling bubbles.
Leveling ratchet wrench.
Lifting jacks.
Light switch.
Mechanical pointers.
Oil filling hole.
Oil filling valve.
Oil screw filler bracket.
Operating cam.
Pedestal spade.
Piston rod.
Recoil cylinder.
Recuperator a i r f i l l i n g valve.
Release lever (fuze setter).
Sliding covers (on indicators).
Squared shaft (temp. adj.).
Top carriage.
Traversing handwheel.
Trunnions.

A. Practical demonstration on mount. See also figures 19 and 20.

14. Action and minor adjustments, mobile mount.—*Q.* Where are the recoil and counterrecoil cylinders located? *A.* All three cylinders are in the cradle under the gun. The center cylinder is the recoil cylinder. The long, small cylinder on the forward end of it is the counterrecoil buffer cylinder. The right-hand cylinder is the air reservoir. The left-hand cylinder is the floating piston cylinder.

Q. Describe the process of recoil. *A.* The three cylinders are connected. The recoil piston comes back when the gun is fired and forces the oil through a small valve into the floating piston cylinder. This valve is constructed so that the size of the opening is proportional to the oil pressure—the higher the pressure, the larger the opening. At the start of recoil, the valve opens wide due to high initial pressure, but as the pressure falls, the valve gradually closes bringing the gun to a smooth stop.

Q. What force brings the gun back into battery after recoil? *A.* The expansion of the air that has been compressed in the floating piston cylinder and the air reservoir cylinder. The compressed air acts on the floating piston, forcing the oil back into the recoil cylinder against the recoil piston, pushing the gun back into battery.

Q. What is the purpose of the counterrecoil buffer? *A.* To slow up the last portion of the movement of the gun in the counterrecoil, thus permitting the gun to return to battery without shock.

Q. Where is the counterrecoil buffer located? *A.* In the forward end of the recoil cylinder.

Q. What are the recoil cylinders and piston fastened to? *A.* The cylinders are all held in the cradle. The recoil piston is fastened to the piston rod which in turn is fastened to the breech ring of the gun and moves with the gun.

Q. What are the equilibrators for? *A.* The gun is muzzle heavy because the trunnions are set far to the rear. The equilibrators are provided to counterbalance this muzzle preponderance and make the gun easy to elevate.

Q. What type of top carriage does this gun have? *A.* Spindle type, carried on a ball thrust bearing instead of rollers.

Q. What type of fuze setter is used with this gun? *A.* The fuze settter M8.

Q. How do you determine whether the gas in the equilibrators is at proper pressure? *A.* By operating the elevating handwheel. If elevation is easy and depression hard the gas pressure is too great and vice versa.

Q. How can the equilibrators be adjusted? *A.* By replenishing the gas supply. This should be done by ordnance personnel.

Q. How is the oil reserve checked? *A.* By bleeding the gun and replacing the oil reserve which has been bled out.

Q. How is the oil reserve bled? *A.* With the gun in a horizontal position, remove the oil filling plug at the lower left rear of the cradle. Insert the oil release tool, and screw it in until the oil filling valve is unseated. If any reserve oil is in the system, it will be forced out through the oil release by the action of the gas on the floating piston.

Q. How is the oil reserve reestablished? *A.* If no oil flows through the oil release, an oil reserve must be established by the injection of oil into the recuperator. Remove the plug from the oil filling hole located on the left side of the cradle (visible through a small hole in the top carriage), and pour in heavy recoil oil through a funnel until the cylinder is completely full. Replace the plug. The proper oil reserve is then established by injecting two oil screw fillerfuls of oil. If oil flowed through the oil release when the oil release tool was applied, only the oil reserve needs to be put in the cylinder.

To inject the oil reserve, fill the oil screw filler and mount in the bracket provided at the rear left side of the cradle. Connect the tube of the filler to the oil filling inlet. Turn the handle of the oil screw filler, forcing the oil into the cylinder. When the handle has been completely turned down, remove the oil screw filler and repeat the operation. Replace the oil filling plug.

Q. What kind of oil is used in the recoil mechanism? *A.* Heavy recoil oil.

Q. How is the air pressure in the counterrecoil cylinders tested? *A.* With the gun in a horizontal position, force the gun out of battery about 1 inch with the jacking device, using the wrench provided. Upon release of the jack, the gun should follow into battery. Elevate the gun to 1.420 mils and repeat the jacking operation. If the gun does not return to battery when the jack has been released, depress slowly and note the angle at which the gun does return to battery. If this angle is below 900 mils, the gas supply must be replenished.

15. Care of gun and mount.—*Q.* What care should be given the bore of the gun after firing? *A.* As soon as possible after firing, it is important that the bore be cleaned to remove all powder fouling and then thoroughly oiled. Using the sponge, wash the bore with a solution made by dissolving ½ to 1 pound of soda ash (depending on strength desired) to 1 gallon of boiling water. A piece of burlap doubled over the bell of the rammer may prove more satisfactory than the sponge. Special attention should be given that portion of the bore extending from the origin of the rifling to a point about 24 inches forward, as most of the fouling takes place in that area. Cleaning should be followed by thorough drying, after which the bore is oiled. A coating of grease should be applied if the gun is not to be fired immediately.

Q. What are some of the routine checks that should be made daily?
A. (1) Open and close the breech to see that it operates freely.

(2) Examine the breech recess and bore to see that they are clean.

(3) See that the firing mechanism works freely.

(4) Elevate and depress the gun to see that the mechanism operates without binding or undue lost motion.

(5) Traverse the gun to the right and left through the full extent of its travel to see that the mechanism operates without binding or undue lost motion.

(6) See that the sliding surfaces of the gun and cradle are clean and well lubricated.

(7) See that all working parts are thoroughly lubricated.

(8) Examine recoil system for oil leaks.

(9) Examine all keys, thongs, and hinges to see that they are in serviceable condition.

(10) Check tools and accessories to see that they are in their proper position and that none is missing.

Q. Of what does maintenance of guns consist? *A.* Routine checks as stated in the answer to the previous question, lubrication at periodic intervals, and cleaning and painting.

Q. Where are cleaning and preserving materials obtained? *A.* They are issued to the battery by the regimental supply officer, on the basis of an allowance prescribed in Ordnance Standard Nomenclature Lists or Army Regulations. Supplies are usually drawn quarterly.

Q. What care should be given the breech mechanism? *A.*

(1) The breech mechanism should be kept clean and well lubricated at all times with class D lubricating oil, light. The mechanism should be disassembled periodically (and always immediately after firing) and cleaned and oiled. In case the mechanism is to be left unused for a considerable length of time, all bright surfaces should be coated with rust-preventive compound.

(2) Vigilance must be maintained to detect any abrasions forming on the pressure side of the wearing surfaces in the various grooves of the breechblock and the breech recess and on the trunnions of the extractors. The removal of such abrasions must be done at once by ordnance personnel.

Q. What care should be given to the firing mechanism?   *A*.

(1) The firing mechanism should be disassembled frequently from the breechblock for the purpose of cleaning and for oiling with light lubricating oil.

(2) The use of an oil that is thicker than authorized will cause the mechanism to absorb the energy of the firing spring and result in misfires. This is especially true in cold weather when unsuitable oil congeals and becomes gummy.

Section III

37-MM GUN

16. Nomenclature.—*Q*. Point out the various parts shown in figure 21.   *A*. See figure 21.

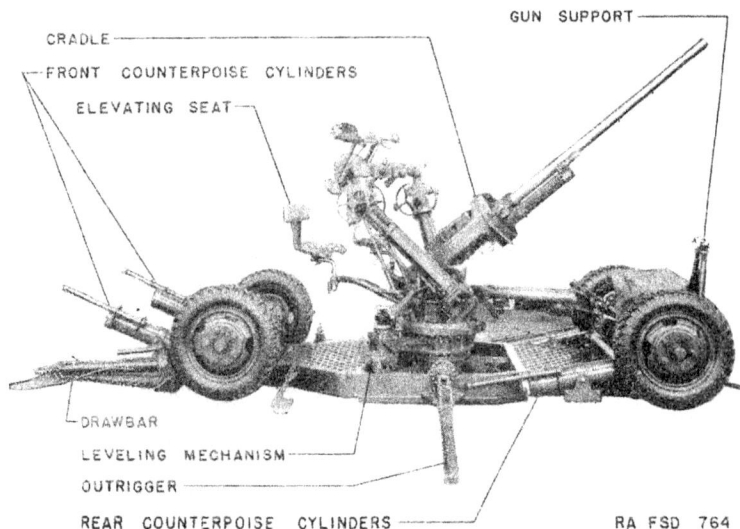

FIGURE 21.—37-mm gun in firing position.

Q. Point out the various parts shown in figure 22. *A.* See figure 22.

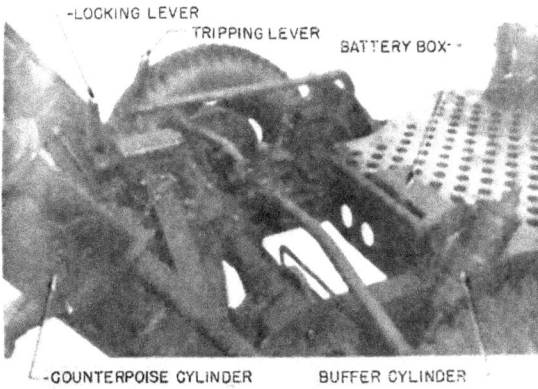

FIGURE 22.—Front of 37-mm gun carriage.

Q. Point out the various parts shown in figure 23. *A.* See figure 23.

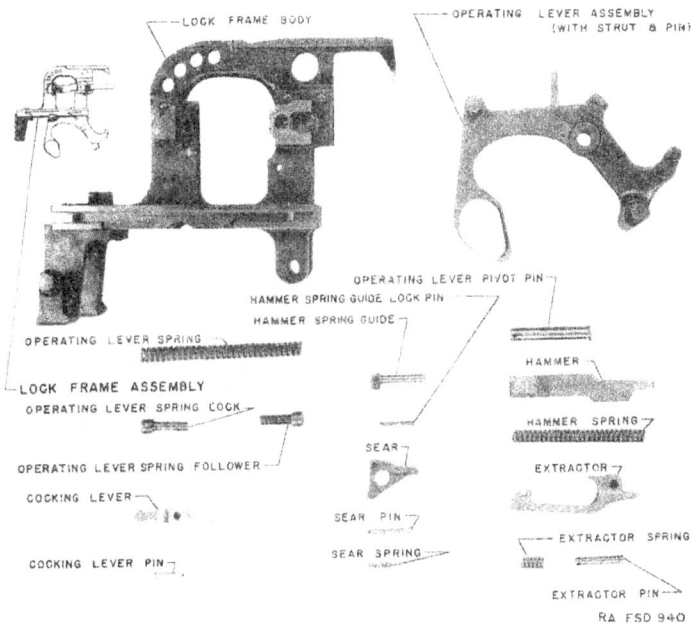

FIGURE 23.—Lock frame.

Q. Point out the parts shown in figure 24.   A. See figure 24.

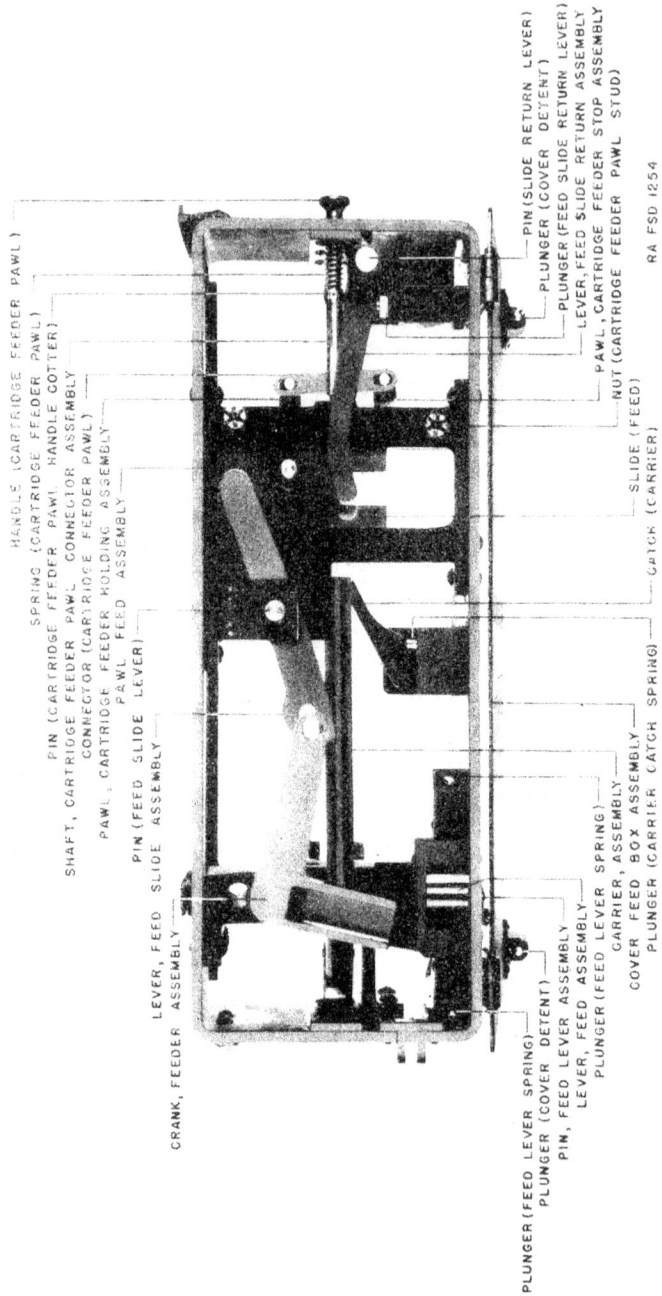

Figure 24.—Feed box.

Labels (reading around the figure):

- HANDLE (CARTRIDGE FEEDER PAWL)
- SPRING (CARTRIDGE FEEDER PAWL)
- PIN (CARTRIDGE FEEDER PAWL HANDLE COTTER)
- SHAFT, CARTRIDGE FEEDER PAWL CONNECTOR ASSEMBLY
- CONNECTOR (CARTRIDGE FEEDER PAWL)
- PAWL, CARTRIDGE FEEDER HOLDING ASSEMBLY
- PAWL FEED ASSEMBLY
- PIN (FEED SLIDE LEVER)
- LEVER, FEED SLIDE ASSEMBLY
- CRANK, FEEDER ASSEMBLY
- PLUNGER (FEED LEVER SPRING)
- PLUNGER (COVER DETENT)
- PIN, FEED LEVER ASSEMBLY
- LEVER, FEED ASSEMBLY
- PLUNGER (FEED LEVER SPRING)
- CARRIER, ASSEMBLY
- COVER FEED BOX ASSEMBLY
- PLUNGER (CARRIER CATCH SPRING)
- CATCH (CARRIER)
- SLIDE (FEED)
- NUT (CARTRIDGE FEEDER PAWL STUD)
- PAWL, CARTRIDGE FEEDER STOP ASSEMBLY
- LEVER, FEED SLIDE RETURN LEVER
- PLUNGER (FEED SLIDE RETURN LEVER)
- PLUNGER (COVER DETENT)
- PIN (SLIDE RETURN LEVER)
- RA FSD 1254

Q. Point out the parts shown in figures 25 and 26.   *A.* See figures 25 and 26.

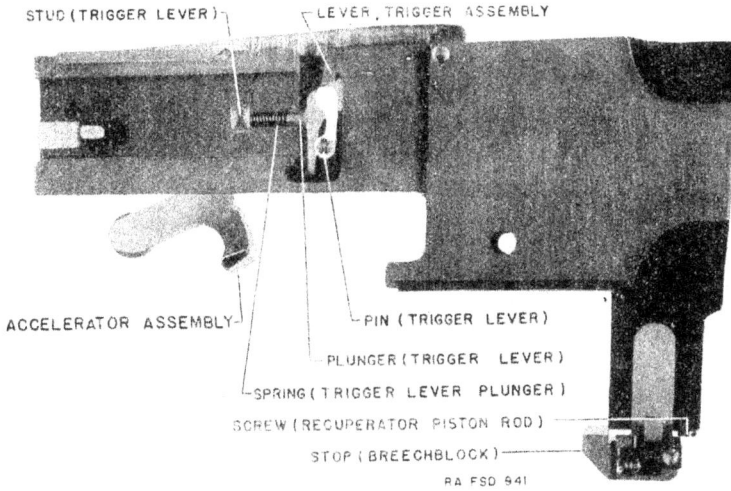

FIGURE 25.—Tube extension (left side view).

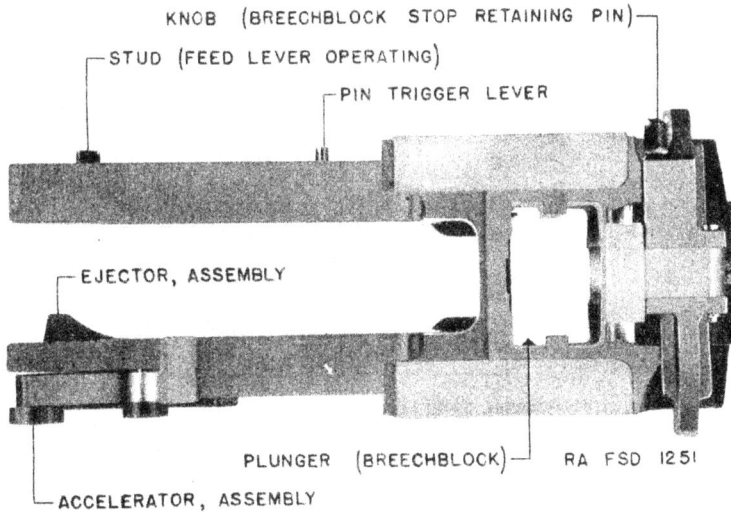

FIGURE 26.—Tube extension (bottom view).

Q. Point out the parts shown in figure 27. *A.* See figure 27.

FIGURE 27.—Leveling mechanism.

Q. Point out the parts shown in figure 28. *A.* See figure 28.

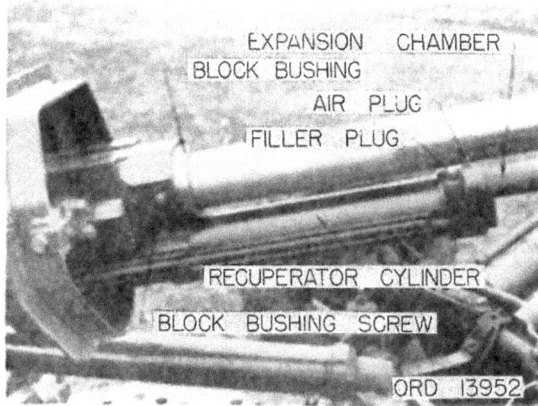

FIGURE 28.—Recuperator.

17. Action and care.—*Q.* What is the weight of the gun and carriage? *A.* 5,365 pounds.

Q. How is the ammunition fed to the gun? *A.* In ammunition clips holding 10 rounds each. See figure 29.

Q. How is an ammunition clip loaded? *A.* Place the clip on a flat surface, prongs up. Place one round at a time on the clip so that the groove in the rotating band is in line with the side wall projections. With one hand on the projectile and the other on the case, press the round straight down into place.

Q. What is the maximum rate of fire? *A.* 120 rounds per minute.

Q. What type of projectile is used? *A.* A high explosive shell equipped with a supersensitive fuze and a tracer element. When the tracer element burns through, it destroys the projectile. For target practice a projectile with a tracer element and an inert filler is used.

Q. What is the approximate weight of the projectiles M54 (HE) and M55A1 (practice)? *A.* 1⅓ pounds.

Q. What is the muzzle velocity of the 37-mm gun when using the standard high explosive or target practice ammunition? *A.* 2,600 feet per second.

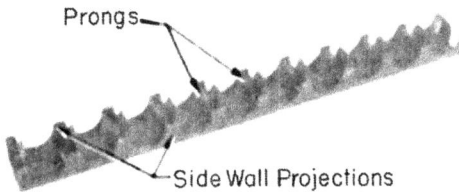

Prongs

Side Wall Projections

Figure 29.—Ammunition clip.

Q. What is the approximate tracer burn-out point? *A.* 3,500 yards.

Q. What are the limits of traverse and elevation of the gun? *A.* 360° traverse and 0° to 85° elevation.

Q. On how great a slope can the gun (emplaced) be leveled by the leveling mechanism? *A.* On a slope not exceeding 5°.

Q. How is the length of recoil measured? *A.* By putting a film of grease on the barrel for 12 or more inches forward of the trunnion block bushing, firing the gun, and then measuring the distance that the grease has been pushed forward by the bushing. This distance is critical. It should be 10¾ inches.

Note.—Before checking the length of recoil at the start of firing, the amount of oil in the recuperator cylinder must be checked by elevating the gun to 5° and, with the filler plug and air plug removed, seeing that the oil is just visible in the bottom of the recess in the bottom of the filler plug of the hole.

Q. How is the length of recoil adjusted? *A.* If the length of recoil is too great, elevate the gun slightly above 5° and put a few additional drops of oil in the recuperator cylinder through the filler plug hole (fig. 28). If the length of recoil is too small, depress the gun slightly below 5° and remove a few drops of oil. After adjustment, fire another round and recheck the length of recoil.

NOTE.—The air plug must be removed from the expansion chamber during the process of filling or extracting oil from the recuperator.

Q. How is the gun cooled? *A.* During firing, the gun is air cooled. This cooling is not sufficient to keep the gun from overheating. The gun must be water cooled after about 50 rounds of

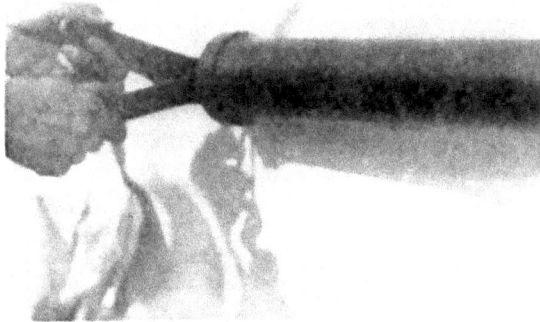

FIGURE 30.—Adjusting equilibrator spring, 37-mm gun.

sustained fire at maximum rate. When single shots or bursts of less than 30 rounds are fired, the gun should be cooled after about 100 rounds. Never fire more than 100 rounds before cooling.

Q. Explain the method of cooling the gun tube with water. *A.* Elevate gun to about 30°. Pull lock frame to rear until caught by carrier dog. Insert the brass tube of the water delivery hose into the breech end of the gun tube and then ease the lock frame forward to hold the brass tube in place. Pump water slowly through the gun tube until the steam generated by the heated gun tube is replaced by a trickle of water from the muzzle. Attach the water return hose to the muzzle of the gun by means of the brass clamping attachment. Pump water through the gun tube for 2 or 3 minutes. Pumping is continued until a man can place his bare hand lightly on the gun tube without danger of being burned. When the gun has been cooled, pull the lock frame to the rear, withdraw the brass tube of the water delivery hose, and allow the water to drain from the gun tube. Remove the water return hose from the muzzle by releasing the clamping attachment. Coil both hose and place them on top of the water chest.

NOTE.—Wipe the gun tube thoroughly dry before resuming fire.

Q. How is the equilibrator spring adjusted? *A.* Adjustment is made by increasing or decreasing the tension of the spring by turning the nuts on the equilibrator rod. Two special wrenches are provided for this purpose. The tension is varied until the elevating mechanism operates smoothly (fig. 30).

Q. What are the general precautions to be observed in disassembling and assembling the gun? *A.*

(1) Use the proper tool to avoid damaging the part.

(2) Do not remove parts except for essential training, for care and lubrication, or for necessary replacement of parts.

(3) Parts not specifically mentioned in this section should not be removed except by thoroughly qualified personnel.

Q. Explain how to remove and replace the gun tube. *A.*

(1) *To remove.*—Have the gun tube approximately horizontal. Unscrew the two tube bushing screws located on the sides of the block at its front end. Pull the lock frame to the rear until it is held back by the carrier dog. Place the centering device (if available, see note 1) on the recoil mechanism. Place the tube wrench in the slots in the gun tube provided for it. Push in the tube latch lock located in front of the feed box and hold the latch in until the tube is rotated counterclockwise (as seen from the muzzle) about 1 inch with the tube wrench. The latch lock can then normally be released. However, if rotation of the tube against the lock in firing has caused the lock to stick, tap the lock lightly until it is disengaged from the tube. Turn the tube counterclockwise (see note 2) until the threads are disengaged from the tube extension and then pull the tube forward out of the gun. The tube bushing will be removed by the rear shoulder on the tube. Care must be exercised to guide the tube as it leaves the trunnion block to prevent it from falling. Place the tube on a canvas cover or length of burlap stretched out on a flat surface to avoid danger of scratching the tube.

(2) *To replace.*—Have gun tube approximately horizontal and lock frame in rear position. Place the centering device (see note 1) on the recoil mechanism. To reduce galling of the threads of the

tube, coat them with a mixture of graphite and castor oil or, in the absence of these, with lubricating oil. Guide the threaded end of the tube into the trunnion block until it is pressing lightly against the threads of the block. Maintain this pressure and rotate the tube counterclockwise until a distinct click is heard signifying that the end of the thread on the tube has passed the end of the thread in the trunnion block. Screw in the tube by hand until it is seated (see note 2). When the tube is properly seated, the latch lock will automatically engage the notch in the tube.

NOTE.—1. If a centering device is not available, a member of the gun section must support the muzzle of the tube at the proper height to prevent the weight of the tube exerting a binding force on the threads.

2. In removing the tube, after the tube has been given about one-quarter turn it should unscrew without the aid of the tube wrench. Similarly in replacing the tube it may be necessary to use the tube wrench to make the last quarter turn.

Q. Explain how to remove and replace the breechblock. *A.*

(1) *To remove.*—Pull the lock frame to the rear until it is held back by the carrier dog. Pull out the spring plunger at the lower right side of the tube extension, which releases the breechblock stop. The breechblock will then drop out. If difficulty is experienced in pulling out the plunger, push the breechblock up a little from the stop.

(2) *To replace.*—Reverse the operations given above. Be sure that the breechblock is resting on the breechblock stop before letting the lock frame go forward.

Q. Explain how to remove and replace the lock frame. *A.* See figure 23.

(1) *To remove.*—Have the gun at a low elevation. See that the lock frame is in its forward position. Push the carrier catch to the right, allowing the carrier to spring downward. Pull the locking clips of the driving spring rod hooks outward and downward and disengage the hooks from the lock frame trunnions. Remove the backplate by pushing forward on the backplate latch located on the bottom of the right-hand side plate and then forcing the backplate downward. Remove the lock frame from the gun, keeping a firm grip on the operating lever and the top of the lock frame to prevent the operating lever from springing forward and injuring the operator. When the lock frame has been removed from the gun carefully swing the operating lever forward.

(2) *To replace.*—Reverse the operations given above. Before reassembling the lock frame be sure that the carrier is down.

Q. Explain how to remove and replace the feed mechanism. *A.* See figure 24.

(1) *To remove.*—Raise and lay back the feed box cover. Withdraw the feed slide lever pin. Force the feed slide to the right and lift out the feed slide lever. Remove the feed lever pin from the right side of the receiver and lift out the feed lever. Remove the feed crank. Remove cotter pin from slide return lever pin and remove pin. Then remove feed slide return lever assembly. Push in on the feed pawl pin with a match or stiff wire and remove the pin. Remove the pawl and pawl spring. Withdraw the slide.

(2) *To replace.*—Reverse the operations given above.

Q. Explain how to remove and replace the carrier. *A.*

(1) *To remove.*—Rotate the carrier pin slightly to the front and withdraw it from the left side of the receiver. Withdraw the carrier to the rear through the bottom of the feed box and out through the backplate opening.

(2) *To replace.*—Reverse the operations given above.

Q. Explain how to remove and replace the carrier catch. *A.*

(1) *To remove.*—Push in on the carrier catch spring plunger until it is clear of the catch shoulder. Remove the catch and carefully release the pressure on the plunger to prevent the plunger and spring from flying out.

(2) *To replace.*—Reverse the operations given above.

Q. Explain how to remove and replace the tube extension. *A.* See figures 25 and 26.

(1) *To remove.*—Remove in order the backplate, lock frame, driving spring assemblies, breechblock, feed mechanism, carrier, tube, and accelerator cam. With the gun in the horizontal position, unscrew recuperator piston rod nut, using the wrench provided, and pull the tube extension out to the rear. *Do not have the gun elevated while piston rod nut is being unscrewed and the tube extension removed.*

(2) *To replace.*—Reverse the operations given above. The gun must be horizontal. The recuperator piston rod nut is screwed in until its outer face is just flush with the inner face of the lip of the breechblock stop when the latter is in the raised position, and the two short edges of the nut are parallel to the sides of the receiver.

Q. Explain how to disassemble and assemble the lock frame. *A.* See figure 23.

(1) *To disassemble.*—Insert a screw driver in the slot in the operating lever spring head at the rear of the lock frame, rotate the head about one-quarter turn counterclockwise, and withdraw the operating lever spring, head, and follower. No further stripping of the lock frame should be done unless necessary for replacement of parts or periodic cleaning and lubrication.

(*a*) To remove the operating lever withdraw cotter pin from the operating lever pin and remove the operating lever pin and operating lever.

(*b*) To remove the extractor, withdraw cotter pin from the extractor pin and remove the extractor pin and extractor.

(*c*) To remove the sear, withdraw the cotter pin from the sear pin and remove the sear pin and sear.

(*d*) To remove the hammer and cocking lever, withdraw the cotter pins from the hammer pin and cocking lever pin. Remove the hammer pin and cocking lever pin. Withdraw the hammer and cocking lever.

(2) *To assemble.*—Reverse operations given above. After the cotter pins are inserted, each one must be bent around the corresponding pin so that it does not interfere with the operation of the gun.

Q. Explain how to disassemble and assemble the tube extension. *A.* See figures 25 and 26.

(1) *To disassemble.*—After removing the tube extension from the gun, proceed as described below.

(*a*) To remove the accelerator, push in on the accelerator spring plunger and remove the accelerator. Then remove the plunger and spring.

(*b*) To remove the trigger assembly, pull out on the trigger trip pin and remove the trigger trip. Push in on the trigger level spring plunger, remove the trigger lever pin, and remove the trigger lever and trigger lever connector. Remove the plunger and spring.

(*c*) To remove the ejector, withdraw the cotter pin from the ejector stud nut and remove the nut, ejector, and the ejector spring. Be careful that ejector spring does not fly out while ejector is being removed.

(*d*) To remove the breechblock stop, withdraw the cotter pin from the breechblock stop pin and remove the stop pin and breechblock stop.

(*e*) Do not perform the operations in (*c*) and (*d*) unless it is necessary for replacement of parts or for periodic cleaning and lubrication.

(2) *To assemble.*—Reverse the operations described above.

Q. In general what care must be taken of the gun? *A.* The gun and mount must be kept in operating condition at all times. Parts must be kept clean and unpainted surfaces covered with a light coating of oil. Instructions for lubrication must be carefully carried out. Covers must be used whenever the gun and mount are not in use and are exposed to the weather.

Q. How should the bore be cleaned? *A.* The bore is cleaned with hot water and issue soap, a sal soda solution, or in the absence of these, hot water. Insert the bore brush in the muzzle. Elevate the gun slightly. pour some of the cleaning solution in the muzzle, and then work the brush back and forth the length of the barrel several times. Flush the bore with clear water. Examine the bore, and if it is not clean, repeat the cleaning operation. When the bore is clean, place burlap or cotton rags on the end of the rammer staff and work the staff back and forth in the barrel until the bore is dry. When the bore is thoroughly dried, apply light lubricating oil. class D, by means of a brush. cotton rags. or burlap attached to the rammer staff.

Q. How is the sal soda solution prepared? A. It is prepared by dissolving ½ to 1 pound. depending upon the strength desired, of soda ash (sodium carbonate cleaning compound—calcined soda) in 1 gallon of boiling water. The solution is prepared at the time it is to be used.

Q. In addition to cooling and cleaning the bore, what other care of the gun is essential? A. All movable parts of the loading and firing mechanisms housed within the feed box and receiver are removed, carefully cleaned. lightly oiled. and then reassembled.

Q. How is the operation of the "breakaway" switch tested? A. By a trial application. The breakaway switch is turned to "on" while the trailer is being towed slowly. If the brakes do not operate properly replace the battery on the trailer and retest.

18. Safety precautions.—Q. What is the purpose of the safety precautions discussed in this paragraph? A. They are prescribed for use in time of peace to insure the safety of the towing airplane and the personnel on the firing line. The fundamentals indicated should be applied under war conditions where circumstances permit.

Q. When may a gun be loaded? A. Only when the command to load has been given by the officer conducting the firing. Guns will always be unloaded except when firing or about to fire.

Q. What precaution should be taken to prevent live ammunition from being used inadvertently in the gun? A. No live ammunition should be allowed near the emplacements except when firing is to take place.

Q. When can persons go in front of the gun? A. No person will be allowed to go in front of the firing line until permission has been granted by an officer, who has ordered all guns to be cleared.

NOTE.—Members of the gun section should always pass in rear of the gun when going from one side to the other.

Q. What precautions are taken with relation to the pointing of the gun? A.

(1) The gun is always kept pointed inside the limiting (left and right) boundaries of the field of fire except when boresighting or other

operations require it to be pointed in some other direction. Before the gun is pointed outside the limits of the field of fire, it must be carefully checked to insure that it is unloaded.

(2) A loaded gun must always be pointed at a safe part of the field of fire.

(3) The gun is depressed considerably below the elevation of the towing airplane until the airplane has passed the line of sight of the gun sights.

Q. When beginning to track a target, at what time does the gunner first place his foot on the trigger pedal? *A.* When his sight is alined with the target.

Q. What is done if the towing plane develops engine trouble? *A.* CEASE FIRING is executed immediately.

Q. When firing ceases at the firing point, what precautions are taken immediately? *A.* The gun is unloaded.

Q. What precautions must be taken to protect against a dragging towline? *A.* When targets are towed over or close to gun positions, overhead cover of some sort must be available so that personnel will be protected from a dragging towline.

SECTION IV

MACHINE GUNS

19. General characteristics.—*Q.* What is a machine gun? *A.* A weapon which fires small-arms ammunition automatically.

Q. What is the name of the machine gun in use in antiaircraft units which fires caliber .30 ammunition? *A.* Browning machine gun, caliber .30, M1917.

Q. What is the standard machine gun? *A.* The Browning machine gun, caliber .50, M2.

Q. What is the present standard machine-gun mount? *A.* The M2 pedestal mount.

Q. Is this mount used for both caliber .30 and caliber .50 machine guns? *A.* Yes. However, a special subcradle or adapter must be used for the caliber .30 gun.

Q. What other mount is sometimes employed for both caliber .30 and caliber .50 machine guns? *A.* The M1 tripod mount.

Q. Describe machine guns as to their operation, ammunition feed, and cooling. *A.* They are recoil-operated, belt-fed, and water-cooled.

Q. Compare the caliber .50 machine gun with the caliber .30 machine gun. *A.* Mechanically they are very similar. The main differences are:

(1) The caliber .50 gun has an oil buffer in place of the lock frame.

(2) The extractor feed cam, just back of the extractor cam, is pivoted, actuated by a spring, and called the switch in the caliber .50 gun.

(3) The caliber .50 gun has a water circulating system for cooling the gun while the caliber .30 gun depends on the water in the water jacket to cool the gun properly.

(4) The caliber .50 gun is provided with a side-plate trigger to permit the gun to be fired by depressing a lever on the M2 mount. The gun may also be fired by depressing the butterfly trigger in the backplate.

(5) The caliber .50 gun may be assembled to fire either with right-hand or left-hand feed.

Q. What are the approximate weights of the caliber .30 and caliber .50 machine guns when the water jackets are filled with water? *A.*

(1) *Caliber .30, M1917.*—41 pounds.

(2) *Caliber .50, M2, 36-inch.*—110 pounds.

(3) *Caliber .50, M2, 45-inch.*—122 pounds.

Q. How much water is required to fill the water jackets of these guns? *A.*

(1) *Caliber .30, M1917.*—7 pints.

(2) *Caliber .50, M2, 36-inch.*—8 quarts.

(3) *Caliber .50, M2, 45-inch.*—10 quarts.

NOTE.—In the case of the caliber .50 gun an additional 8 gallons (approximate) is required for the water chest.

Q. What is the approximate weight of ammunition chests containing filled belts? *A.*

(1) *Caliber .30, web belt (250 rounds).*—21 pounds.

(2) *Caliber .50, metallic link belt (200 rounds).*—89 pounds.

Q. What is the muzzle velocity of the caliber .30, M1917, machine gun? *A.* 2,700 feet per second.

Q. What is the muzzle velocity of the caliber .50, M2, machine gun? *A.* 2,700 feet per second.

Q. What are the rates of fire of the caliber .30 and caliber .50 machine guns? *A.*

(1) *Caliber .30.*—400 to 525 shots per minute.

(2) *Caliber .50.*—500 to 650 shots per minute.

NOTE.—These are the rates for a short burst of continuous fire. The usable rates for sustained fire consisting of a series of bursts will be much lower.

Q. What means is provided to help the gunner steady the gun in firing? *A.* A back rest or shoulder stock. The back rest is preferable.

20. Nomenclature, action, and care.—*a. General.*—(1) *Nomenclature.*—*Q.* What are the groups or assemblies of which the machine gun consists? *A.* The bolt group, oil-buffer group (lock frame group in caliber .30 gun), barrel group, cover group, backplate group, and casing and water jacket group.

(2) *Action.*—*Q.* What does the belt feed mechanism do? *A.* As soon as a cartridge is withdrawn it feeds the belt forward one loop and puts the next cartridge in place to be withdrawn.

Q. What does the extractor do? *A.* As the bolt comes back the extractor pulls a cartridge out of the belt and drops it into the T-slot in the face of the breech, the ejector on the end of the extractor having knocked the empty case out of the T-slot ahead of the full cartridge. As the bolt goes forward the extractor rises to grip the next cartridge in the belt while the bolt pushes the preceding one into the chamber.

Q. What is the purpose of the breech locking mechanism? *A.* To keep the breech closed tightly after the shot is fired until it is safe to open it.

Q. How is this done? *A.* By means of the breech lock, which locks the bolt to the barrel extension and does not release it until the barrel and bolt assembly have recoiled together a short distance.

Q. What are the main parts of the firing mechanism? *A.* The trigger, sear, sear slide (caliber .50), sear spring, firing pin and extension, and cocking lever.

Q. How does the firing mechanism work? *A.* When the trigger is operated, the sear is tripped allowing the firing pin to go forward and strike the rear end of the cartridge, firing the gun. Actuated by the shock of firing, the bolt moves to the rear, recocking the gun. If the gunner continues to hold the trigger in the firing position, the gun will fire automatically each time the bolt returns to its forward position.

Q. What is the purpose of the buffer mechanism? *A.* To cushion the blow of the bolt when it is stopped at the end of its recoil.

Q. What are the main parts of the buffer mechanism of the caliber .30 machine gun? *A.* Grip, buffer plate, buffer ring (which is split on one side), a number of fiber washers (buffer disks), and the adjusting screw. The caliber .50 mechanism is similar.

Q. What other shock-absorbing device is there in the caliber .50 gun? *A.* The oil buffer, which corresponds to the lock frame in the caliber .30 gun.

Q. What is its purpose? *A.* To cushion the shock of the recoil of the barrel and bolt assembly before the bolt is unlocked and opened.

It is necessary on the caliber .50 gun because of its high power and heavy recoil.

Q. How can the rate of continuous fire of the caliber .50 gun be regulated? *A.* By inserting a screw driver blade into the slot in the rear of the oil buffer tube. Turn the tube clockwise to reduce the rate of fire, and counterlockwise to increase the rate of fire.

NOTE.—The continuous rate of the caliber .30 gun may be increased by adding one or more buffer disks to the shock-absorbing group in the backplate grip.

Q. What is the purpose of the driving mechanism? *A.* To return the moving parts to the closed position after they have recoiled. The driving spring and the barrel plunger spring are compressed in recoil. As soon as the recoil ends the springs push the parts back into the closed position again.

Q. Why is the cooling system necessary and important? *A.* Because the gun gets very hot if fired continuously. It would be damaged badly if there were no way to cool the barrel. Proper cooling of the gun is also required to reduce erosion of the bore to a minimum. Consequently it is very important to keep enough water in the water jacket.

Q. How is the barrel packed, and why is it necessary? *A.* Special asbestos packing, saturated in oil, is applied to the muzzle and breech ends to prevent leakage of water.

Q. How can steam escape from the water jacket when the water gets hot? *A.* The inner steam escape tube runs lengthwise at the top of the water jacket and has a hole at each end. It has an outer tube sliding freely on it just short enough to uncover one of the end steam holes at a time. If the gun is elevated the outer tube slides back and covers the rear hole, preventing the water from running out, but uncovering the front hole and allowing the steam to escape.

(3) *Adjustment.—Q.* What is head space? *A.* The distance from the face of the bolt to the base of the cartridge when the latter is fully seated in the chamber.

Q. What is the purpose of head space adjustment? *A.* To obtain the proper distance between the forward part of the bolt and the rear end of the barrel. This distance is not the actual head space.

Q. How is the head space adjustment made? *A.* It is made without removing the working parts of the machine gun from the casing. To head space the caliber .50 M2 gun, screw barrel by hand into barrel extension until it comes in contact with the bolt. Check to make sure that the end of the barrel extends through the barrel extension. Then unscrew the barrel two notches. If the gun operates sluggishly, unscrew the barrel one additional notch.

77

Q. What happens if head space is too small (adjustment too tight)? *A.* The breech lock will not fully enter its recess in the bolt. The gun operates sluggishly and the barrel extension, bolt, or breech lock may become damaged.

Q. What happens if head space is too great? *A.* A separation of the cartridge case may occur. If there is any weakness in the head of the cartridge case, such as a split case, the possibility of a rupture is increased by excessive head space.

Q. How is the buffer mechanism in the backplate adjusted? *A.* By increasing or decreasing the number of fiber buffer disks in the grip. When the backplate is reassembled, the adjusting screw must be screwed tightly against the buffer disks.

Q. What is a stoppage? *A.* Any unintentional cessation of fire.

Q. What is immediate action? *A.* The procedure used for the prompt reduction of common stoppages.

Q. What are some possible stoppages? *A.*

(1) Misfire due to defective primer.

(2) Short round.

(3) Bulged round.

(4) Tight link or loop in belt.

(5) Thin rim, permitting nose of bullet to drop below chamber.

(6) Stretched or torn belt (fabric).

(7) Empty loop in belt.

(8) Ammunition improperly alined in belt.

(9) Battered or thick rim of cartridge.

(10) Failure to remove round from chamber.

(11) Set-back primer.

(12) Separated case which is removed from chamber by new round when bolt is pulled to rear.

(13) Separated case which stays in chamber when bolt is pulled to rear.

(14) Bullet loose in cartridge case. Cartridge case extracted from belt but bullet remains in belt.

(15) Short or broken firing pin.

(16) Weak or broken firing pin spring.

(17) Faulty engagement of firing pin and sear notch.

(18) Broken sear spring.

(19) Bent or worn belt feed lever.

(20) Belt feed pawl spring missing or weak.

(21) Belt feed pawl pin missing or out of position.

(22) Cover extractor spring missing or weak.

(23) Belt feed lever bent up (stud on lever jumps out of cam groove).

(24) Broken or damaged extractor.

(25) Belt holding pawl missing or spring weak.

(26) Broken or damaged ejector.

(27) Broken or damaged T-slot causing misalinement and buckling of cartridge as bolt moves forward.

(28) Weak ejector spring, causing misalinement and buckling of cartridge as bolt moves forward.

(29) Broken barrel extension.

(30) Defective trigger mechanism.

(31) Defective bolt switch.

(32) Bent or broken belt feed pawl arm.

(33) Cover not latched.

Q. Describe the procedure for immediate action.   A.

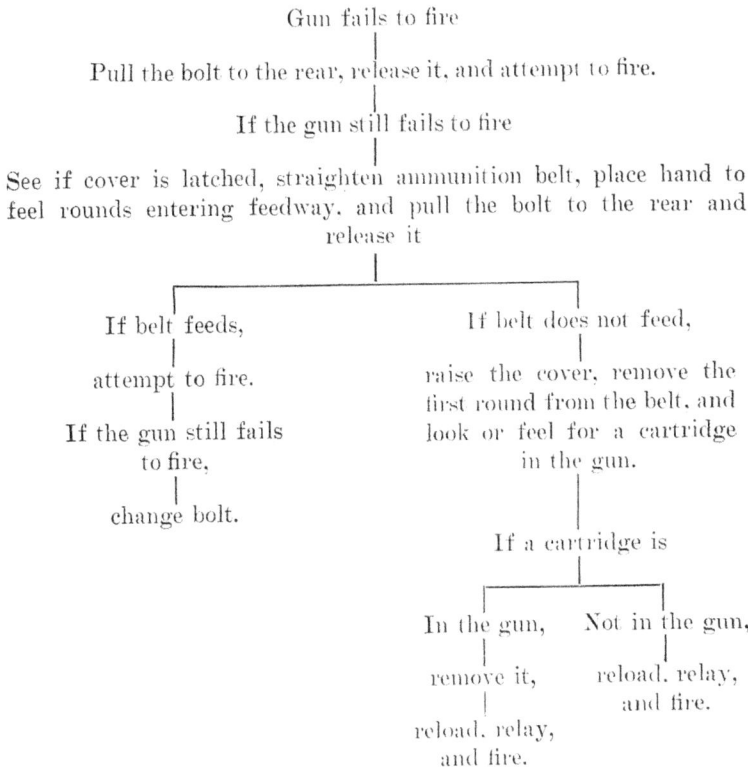

Gun fails to fire
|
Pull the bolt to the rear, release it, and attempt to fire.
|
If the gun still fails to fire
|
See if cover is latched, straighten ammunition belt, place hand to feel rounds entering feedway, and pull the bolt to the rear and release it

If belt feeds,
|
attempt to fire.
|
If the gun still fails to fire,
|
change bolt.

If belt does not feed,
|
raise the cover, remove the first round from the belt, and look or feel for a cartridge in the gun.
|
If a cartridge is

In the gun,
|
remove it,
|
reload, relay, and fire.

Not in the gun,
|
reload, relay, and fire.

NOTE.—If application on this procedure does not remedy the stoppage, the gunner must examine the feed mechanism and other parts of the gun in order to locate and remedy the trouble.

Q. What is the usual procedure in replacing parts during firing periods? *A.* Remove and replace the entire group or assembly of which the broken part is a member. This is particularly applicable to the bolt, which contains most of the parts liable to break.

Q. What is done with the assembly thus removed? *A.* It is disassembled and the broken part is replaced. The group is then reassembled and held as a spare.

(4) *Care.—Q.* On what does the accuracy and proper functioning of a machine gun chiefly depend? *A.* On the proper care and adjustment of the gun and its mount.

Q. What general duties must be performed in caring for the gun and mount? *A.* Cleaning, lubricating, and handling the gun and mount, and the timely replacement of worn or broken parts.

Q. What parts require particular attention when cleaning and lubricating the gun? *A.* The bore, chamber, receiver, and all moving parts.

Q. When should the bore be cleaned? *A.* Immediately before and immediately after firing.

Note.—Inspect and clean the guns daily for 3 days after firings are ended.

Q. What two kinds of barrel fouling are there? *A.* Powder fouling and metal fouling.

Q. What harm does powder fouling do? *A.* The burnt primer and powder ash left in the bore after firing eats into the steel if left for any time. The barrel becomes pitted and inaccurate.

Q. What harm does the metal fouling from the jackets of the bullets do? *A.* Its greatest harm is covering up the powder fouling so that it is not removed, and remains to corrode the bore.

Q. How is powder fouling removed? *A.* Immediately after firing is ended for the day, disassemble the groups from the gun. Insert the barrel, muzzle down, in a vessel containing hot water and issue soap, a sal soda solution, or, lacking either of these, hot water. Insert cleaning rod, with a flannel patch on its end, in the breech end of the bore. Move rod forward and back for about 1 minute, pumping water into and out of the bore. Replace the flannel patch with a brass or bronze wire brush and while the bore is wet, run the brush forward and back through the bore three or four times. Pump water through the bore again using the flannel patch. Dry the cleaning rod and remove the barrel from the water. Using dry, clean flannel patches, thoroughly swab the bore until it is perfectly dry and clean. Thoroughly dry and clean the chamber using a flannel patch on a stick if necessary. Saturate a patch with sperm oil and swab bore and chamber with the patch. Allow a thin coat of the oil to remain in the bore.

Q. What should be done if metal fouling is discovered in the bore of the gun? A. The barrel should be turned in to the ordnance supply agency for removal of the fouling.

Q. What should be done to the working parts after firing? A. All of them should be wiped clean, care being taken to remove dirt and lint from the belt-holding pawl. All parts should then be wiped with an oily rag.

Q. If the gun is to be put away for some time what should be done? A. All metal parts should be thoroughly covered with rust-preventive compound, medium.

Q. What is done to the water jacket after firing? A. It is drained and wiped clean, and the muzzle gland is tightened and repacked if necessary.

Q. What care is taken of ammunition? A. It is watched for corrosion, if opened. Belts should be hung up, clean and dry.

Q. What precautions should be taken in cold weather? A.

(1) See that water in the cooling system does not freeze.

(2) Test the gun frequently to see that it functions properly.

(3) When practicable to do so, keep the movable parts and the receiver absolutely dry and free from oil while the gun is not being fired. However, these parts must be re-oiled before the gun is fired.

Q. State in detail the points to be observed before, during, and after firing. A. See following chart.

POINTS TO BE OBSERVED BEFORE, DURING, AND AFTER FIRING

Points	Before	During	During temporary cessation	After
Bore	Look through and clean.		Clean bore.	Clean and oil.
Muzzle gland	Tighten.		Examine and tighten if required.	Examine and repack if required.
Moving parts	Examine for dirty parts; oil; and test for worn or broken parts. See that parts function without excessive friction.	KEEP OILED.	Inspect and oil. Clean dirt from belt-holding pawl.	Remove bolt, lock frame (oil buffer), barrel extension, and barrel; clean, oil, and replace; and release firing pin.
Head space	Make correct adjustment and test. Examine barrel-locking spring.	Tighten if several separated cases occur.		Adjust correctly and test. Examine barrel-locking spring.
Packing	Examine for leakage or excessive friction.	Watch for leakage.	Examine for leakage.	Examine and replace if necessary.
Tripod or pedestal	Set up firmly, with no lost motion.	Keep firmly set.	Examine.	Clean and oil.
Water	Fill water jacket, see that water plugs and hose connections are tight and that there is a sufficient supply of water. See that water pump functions properly. (See note.)	Keep water jacket and water chest full and plugs and hose connections tight. (See note.)	Keep water jacket and water chest full and plugs and hose connections tight. (See note.)	Empty all water into water chest. Wash out or flush water jacket if muddy or dirty water has been used.
Belts and ammunition.	See that belts are clean. Secure sufficient supply of ammunition. Load belts. Inspect ammunition. Place loaded belts in ammunition chests.	Keep belt in line with feed opening. Replace ammunition chest with new one when necessary. Watch ammunition supply and obtain additional ammunition when necessary.	Replace ammunition chest with full one if necessary. Refill belts in reserve ammunition chests. Check approximate number of rounds fired on course. Replenish ammunition supply if necessary.	Clean, repair, and refill all belts. Separate live rounds from empty cases. Inspect ammunition.
Oil	See that oil can is full.	Keep within reach.	Refill oil can.	Refill oil can.
Spare parts and tools.	Keep clean and oiled. See that kits are complete.		Make repairs. Replace broken or worn parts.	Check; replace broken or missing parts; clean and oil.
Gun record	Check from last firing to see that repairs have been made.		Make record of repairs and gun performance on gun report. Record number of rounds fired.	Complete gun record as indicated under "During temporary cessation."

Note.—If the caliber .50 gun is fired without a circulating unit, the outlet hole in the water jacket must be left uncovered.

b. Caliber .30 machine gun.—(1) *Nomenclature and action.*—*Q.* point out and state the uses of the important parts shown in figure 31. *A.* See figure 31.

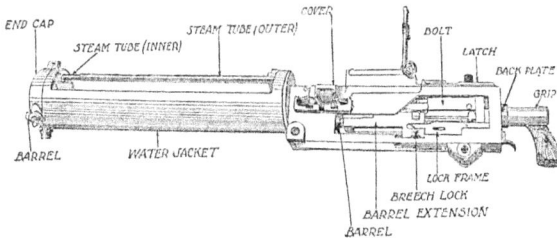

FIGURE 31.—Browning machine gun, caliber .30, M1917.

Q. Point out and state the uses of the important parts shown in figures 32 and 33. *A.* See figures 32 and 33.

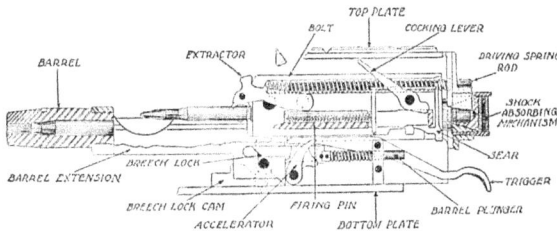

FIGURE 32.—Moving parts of caliber .30 machine gun in rearmost position.

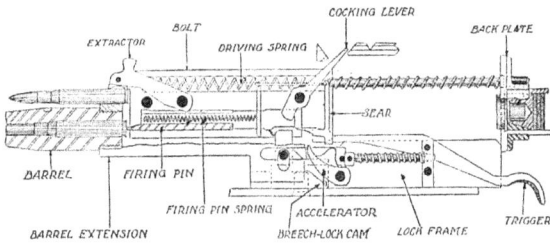

FIGURE 33.—Moving parts of caliber .30 machine gun in forward position.

Q. Point out and state the uses of the important parts shown in figure 34. *A.* See figure 34.

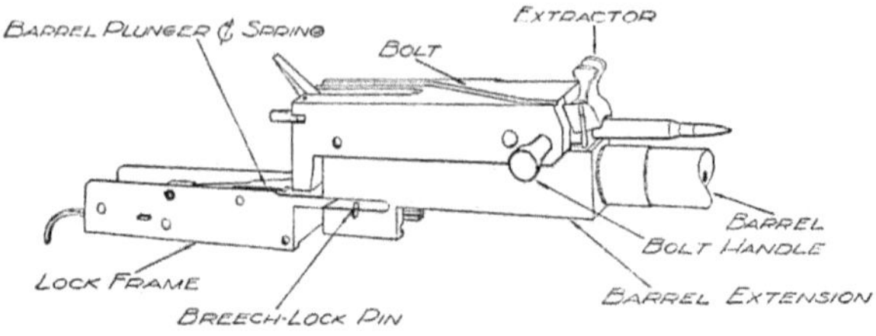

FIGURE 34.—Barrel, barrel extension, bolt, and lock frame in firing position.

Q. Point out and state the uses of the important parts shown in figure 35. *A.* See figure 35.

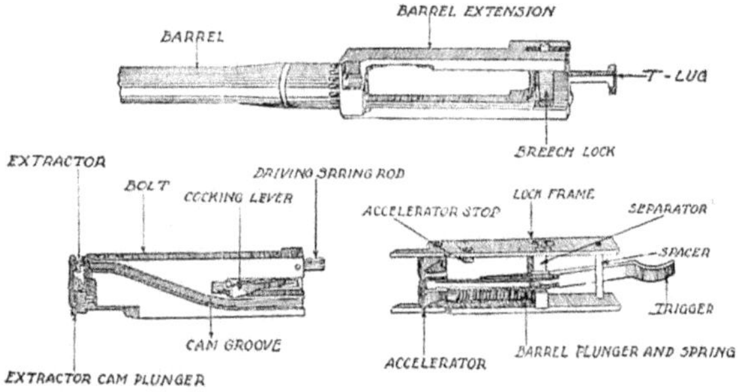

FIGURE 35.—Barrel, bolt, and lock frame groups.

Q. Point out and state the uses of the important parts of the belt feed mechanism of the caliber .30 machine gun. *A.* See figure 36.

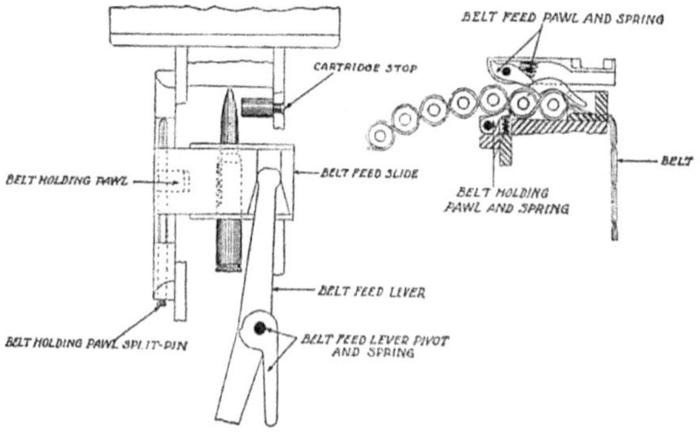

FIGURE 36.—Belt feed mechanism, caliber .30 machine gun.

(2) *Disassembly and assembly.—Q.* Explain and demonstrate the removal and replacement of the groups of the caliber .30 machine gun. *A.*

(1) *Removing groups.—(a) Backplate.*—Pull back on latch and raise cover. With the left hand pull back the bolt handle and hold it in the rearmost position. Insert the rim of a cartridge in the slot in the rear of the driving spring rod. With the slot horizontal, push in the driving spring rod as far as it will go and turn it clockwise one-quarter turn until the slot is vertical so that the lugs on it will engage in the undercut recesses in the bolt. Then push the bolt handle forward about an inch to free the rear end of the driving spring rod from the backplate. Push the latch forward and lift out the backplate.

(*b*) *Bolt.*—Pull the bolt all the way back and remove the bolt handle. Then pull the bolt out of the rear end of the receiver.

(*c*) *Lock frame.*—Push in on the trigger pin through hole on the right side of the receiver with the point of a bullet or the combination tool. Pull lock frame, barrel extension, and barrel out through the rear of the gun until the lower rear lugs on the barrel extension are clear of the casing. Hold the lock frame firmly and push forward on the accelerator with the thumbs. This separates the lock frame from the barrel extension.

(*d*) *Barrel extension and barrel.*—Pull the barrel extension and barrel out to the rear. Unscrew the barrel extension from the barrel.

(*e*) *Cover.*—Remove latch and latch spring. Turn cover pin spring up. Draw the assembled pin out to the right of the casing. Remove the cover.

(2) *Replacing groups.*—The groups are replaced in the reverse order to their removal.

Q. Explain and demonstrate the procedure of removal and replacement of the groups of the caliber .50 machine gun. *A.*

(1) *Removing groups.—(a)* Release cover latch and raise cover.

(b) *Cover.*—The cover may be removed or left assembled to the gun as desired. To remove the cover, withdraw the cotter pin from the cover pin and pull out the cover pin.

(*c*) *Backplate.*—Release the backplate latch lock and latch and lift out the backplate.

(*d*) *Bolt.*—Press forward and away from the side plate on the end of the driving spring rod to release the retaining pin in the head of the rod from the hole in the side plate. Draw the bolt to the rear until the bolt stud can be withdrawn through the opening in the right side of the receiver of the gun. Remove the complete bolt from the rear end of the casing.

(e) *Oil buffer, barrel extension, and barrel.*—Compress the oil-buffer spring lock, using the point of a bullet or the drift on the combination tool inserted through the hole in the right side plate, and remove the oil buffer, barrel extension, and barrel assembly from the rear of the casing. Detach the oil-buffer group from the barrel extension by pressing the accelerator forward.

(2) *Replacing groups.*—The groups are replaced in the reverse order to their removal. Be sure the backplate latch lock is in the unlocked position while the backplate is being replaced.

Q. When replacing the bolt of either a caliber .30 or caliber .50 gun, what precaution must be taken? *A.* Have the cocking lever in the forward position.

Q. Explain how to disassemble and assemble the bolt of a caliber .30 machine gun. *A.*

(1) *To disassemble.*—(*a*) Turn extractor up and remove it to left.

(*b*) To remove driving spring rod, place protruding end of rod on a table or a block of wood. *Great care should be exercised in removing the driving spring rod from the bolt as the force of the driving spring, when released, can easily cause the rod to slip away from the hand and possibly result in serious injury to personnel.* With the bolt firmly grasped by the right hand (palm of hand over top of bolt), press down on and, at the same time, turn the bolt one-fourth turn to the left until the lugs on the rod leave their recess in the bolt. Slowly release pressure on bolt until about 3 inches of rod protrude. With left hand grasp protruding portion of rod and spring; raise bolt from table. Separate rod and spring from bolt with a quick jerk to prevent the spring kinking. Separate rod and spring.

(*c*) Turn top of cocking lever to rear of bolt and withdraw cocking lever pin to left. Lift out cocking lever.

(*d*) Release the firing pin by pushing down on the sear with the point of a bullet. Hold bolt in left hand, front end toward the body, top up, with index finger of left hand beneath and supporting the sear. Push down and to the right on the end of the sear spring with the point of a bullet to seat the spring in the cut in the bolt. This releases the sear which is removed at the bottom of the bolt. Turn sear spring back to left to release it from bolt. Push point of bullet into hole in bottom of bolt to start sear spring pin up. Place end of cocking lever under sear spring and pry down against edge of bolt. Tilt rear end of bolt down and firing pin will fall out.

(2) *To assemble.*—(*a*) Place a firing pin in bolt, striker downward and to the front. Tilt front of bolt downward until end of striker projects through small hole in the front of the bolt. Push sear spring pin back into place, pushing on end of pin and not on spring. Hold bolt in left hand, front end to body, top up. Push end of sear spring down and to right to seat it. Push sear into place from bottom, notched projection toward front of bolt, and hold with first finger of left hand while pressing downward and to the left on the sear spring with a cartridge to engage the end of the sear spring in the sear.

(*b*) Replace cocking lever, rounded nose on lower end toward rear of bolt. Push upper end of lever to rear and insert cocking lever pin in hole in left side of bolt, engaging lever. Cock by pressing forward on cocking lever. Hold cocking lever to rear and press down on sear with a bullet to release firing pin. This operation tests the assembly of the bolt.

(*c*) *The same care should be exercised in assembling the driving spring rod to the bolt that is exercised in removing it.* Place the driving spring on the driving spring rod. With back end of rod resting on table or wooden block, gather as much of spring on rod as can be held compressed by thumb and fingers of left hand. With bolt held securely in right hand, front end of bolt in palm, slip bolt over end of spring. Push downward to compress spring and allow driving spring rod to enter slot in bolt. Turn bolt slowly one-quarter turn to right, until slot in rod is crosswise to slot in bolt. Check to see that slot of driving spring rod is vertical.

(*d*) Replace extractor by inserting its pin in its recess in bolt and swinging it forward into place.

Q. Explain how to disassemble and assemble the lock frame of the caliber .30 machine gun. *A.*

(1) *To disassemble.*—Grasp the head of the trigger pin between the thumb and first finger of the right hand and remove it to the right. If the pin is too tight to permit its removal in this manner it must be drifted out. Do not remove the trigger pin spring except when necessary. Push out the accelerator pin and remove the accelerator. Hold lock frame with left hand, projections upward, slot to left, separator between second and third fingers, first and second fingers gripping barrel-plunger spring. With thumb of right hand, press down and out on barrel plunger to disengage plunger-guide pin. Allow spring, carrying plunger, to rise slowly. Lift out spring and remove from plunger.

(2) *To assemble.*—Assemble barrel-plunger spring to barrel plunger. Hold lock frame with left hand, projections upward, slot to

left, lock frame separator between second and third fingers. Seat end of barrel-plunger spring in recess in lock frame separator, barrel plunger-guide pin facing slot in lock frame, left. Hold spring with first and second fingers of left hand. With thumb of right hand press down on end of barrel plunger until barrel plunger-guide pin can be seated in slot. Care should be taken that the action of the spring does not cause the plunger to slip out of hand. Replace accelerator, tips up, rounded surface to front. Insert accelerator pin, ends of pin flush with sides of frame. Insert trigger from rear, trigger arm under lock frame spacer and over lock frame separator. Insert trigger pin in trigger pin spring if spring has been removed. Insert trigger pin and spring in hole in right side of lock frame.

Q. Explain the procedure of disassembling and assembling the cover of a caliber .30 machine gun. *A.*

(1) *To disassemble.*—Turn belt feed lever pivot spring outward and remove belt feed lever pivot. Remove belt feed lever by drawing backward from the belt feed slide. Remove belt feed slide. Insert point between cover extractor spring and notch in cover extractor cam and, with left thumb over spring, pry spring out to disengage it from cut. Lift spring from its seat against stud.

(2) *To assemble.*—Place forked end of cover extractor spring under stud on cover. Press downward with thumb on other end of spring, pushing toward the stud, and seat projection on spring in undercut recess in cover extractor cam. Replace belt feed slide in its grooves in cover, pawl pointing to right as cover goes on gun. Place front end of belt feed lever in slot of belt feed slide, stud on the lever away from the cover and to the rear. Insert belt feed lever pivot so that spring is at right angles to the cover and the lugs on the pivot pass through the cuts in the cover. Turn the spring inward toward the base of the rear sight until the lugs engage on the under side of the cover and the spring locks in place.

Q. Explain the procedure for disassembling and assembling the buffer group in the backplate of the caliber .30 machine gun. *A.*

(1) *To disassemble.*—With combination tool unscrew adjusting screw and remove it. Remove adjusting screw plunger and spring. Remove buffer disks, buffer plug, buffer ring, and buffer plate.

(2) *To assemble.*—Hold backplate in left hand by stocks, rear end of grip upward. Drop buffer plate into grip with narrower surface downward. Drop in buffer ring (smaller opening downward), and buffer plug (smaller end downward). Replace buffer disks in grip and jar them into place. Insert adjusting screw plunger in adjusting screw plunger spring and return both to their seating in the grip.

Q. Explain how to disassemble and assemble the barrel extension of the caliber .30 machine gun. *A.*

(1) *To disassemble.*—Insert rim of cartridge under front edge of barrel-locking spring and pull it forward. Push out breech lock pin and remove breech lock.

(2) *To assemble.*—Place the breech lock in its slot, *taking care that the double-beveled surface is up and to the front.* Insert the breech lock pin, insuring that both ends of the pin are flush with the sides of the barrel extension. Insert the barrel-locking spring in its seat in the left side of the barrel extension, hook inward, and force home as far as it will go.

Q. Explain how to disassemble and assemble the belt-holding pawl of the caliber .30 machine gun. *A.*

(1) *To disassemble.*—Hold down belt-holding pawl. Withdraw belt-holding pawl split pin to rear and lift out pawl and spring.

(2) *To assemble.*—Replace belt-holding pawl spring in its seat. Replace pawl and hold it in position while replacing split pin.

Q. Explain how to disassemble and assemble the firing pin of the caliber .30 machine gun. *A.*

(1) *To disassemble.*—Drive out firing pin spring pin with combination tool and remove firing pin spring.

(2) *To assemble.*—Replace firing pin spring. force it into place with combination tool, and replace pin.

c. Caliber .50 machine gun.—(1) *Nomenclature and action.*—*Q.* Identify the part groups of the caliber .50 machine gun M2. *A.* See figure 37.

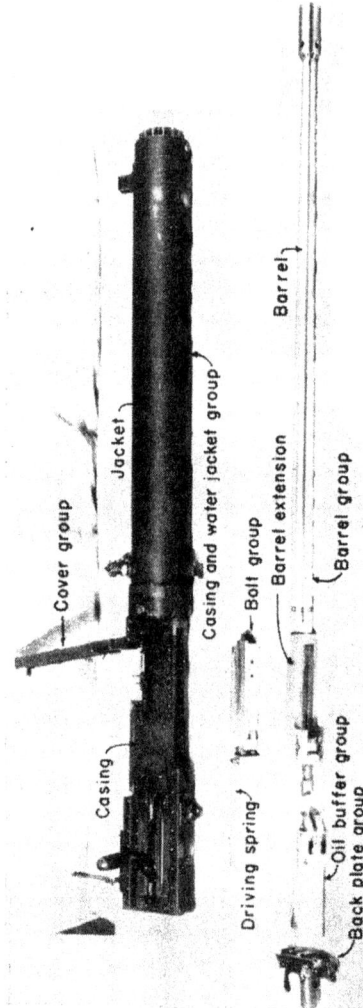

FIGURE 37.—Part groups of caliber .50 machine gun M2.

Q. Identify the component parts of the bolt group. *A.* See figure 38.

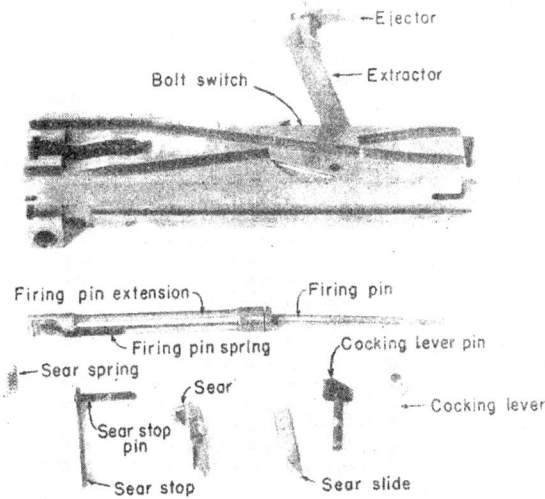

FIGURE 38.—Bolt group.

Q. Identify the component parts of the oil-buffer group. *A.* See figure 39.

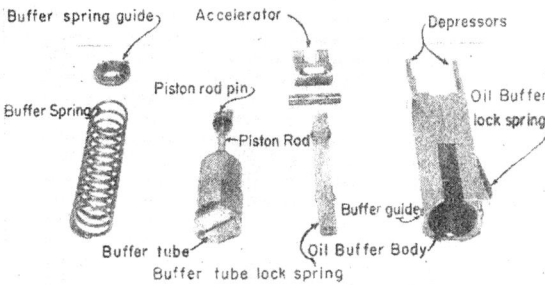

FIGURE 39.—Oil-buffer group.

Q. Identify the component parts of the barrel group. *A.* See
figure 40.

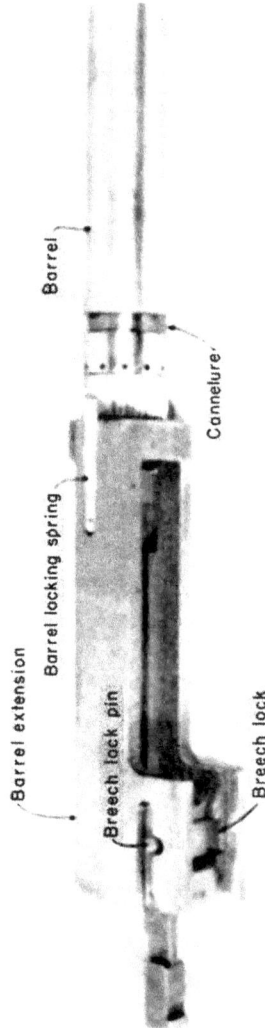

FIGURE 40.—Barrel group.

Q. Identify the component parts of the cover group. *A.* See figure 41.

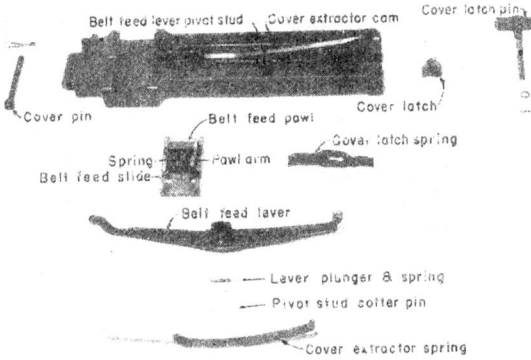

FIGURE 41.—Cover group.

Q. Identify the component parts of the backplate group. *A.* See figure 42.

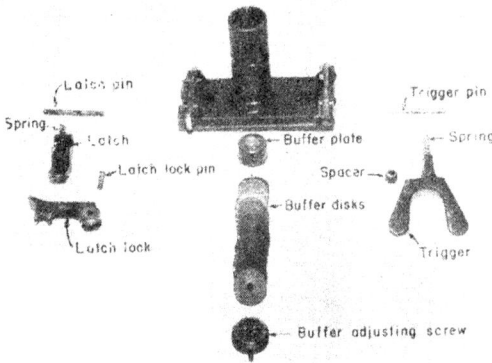

FIGURE 42.—Backplate group.

Q. Identify the component parts of the casing and water jacket group. *A.* See figure 43.

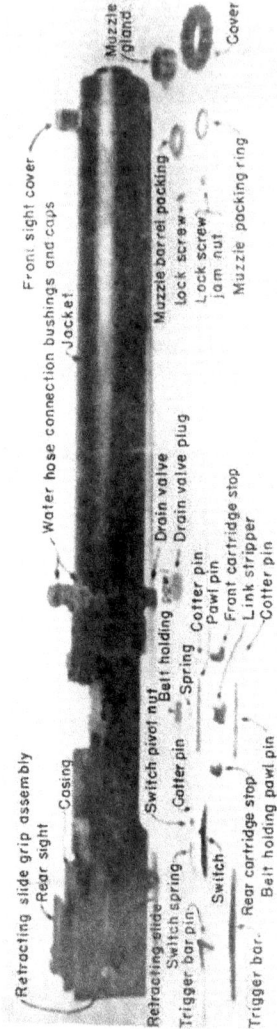

FIGURE 43. Casing and water jacket group.

(2) *Disassembly and assembly.*—*Q*. Explain how to disassemble the bolt. *A*. See figure 38.

(1) The driving spring rod assembly is normally disassembled only when it is necessary to replace the driving spring. To disassemble it, drive out the collar stop pin from the driving spring rod and remove the collar and the driving spring. Next, disassemble the bolt assembly. Remove the extractor by rotating it upward and pulling it out from the bolt. Do not disassemble the ejector from the extractor unless absolutely necessary. To disassemble the ejector, drive out the ejector pin and remove the ejector and ejector spring. Remove the bolt switch from the bolt.

(2) Rotate the cocking lever fully to the rear and release the firing pin by pushing down the sear. Remove the cocking lever pin and cocking lever. With thin end of the cocking lever, push the sear stop to the right until it is in the center of the slot, then turn the bolt over and push the sear stop out of engagement with the firing pin spring. Reverse the bolt and remove the sear stop from the slot. Remove the sear, sear slide, and sear spring. Elevate the front end of the bolt, allowing the firing pin extension and firing pin to slide out. The firing pin spring can be removed from the firing pin extension by driving out the firing pin spring stop pin. Take precautions to prevent firing pin spring from flying out during the operation.

Q. Explain the procedure in disassembling the oil-buffer group. *A*. See figure 39. Drive out the accelerator pin and remove the accelerator. The remainder of the oil-buffer assembly is not disassembled unless absolutely necessary for repairs or replacement. To remove the oil-buffer tube, push in on the end of the oil-buffer piston rod so that the tube may be gripped by hand and pulled to the rear. The oil-buffer spring guide is depressed until it clears the piston rod pin and then is turned until the pin will pass through the slots provided in the guide. Remove the guide and spring.

Q. Explain the disassembly of the barrel group. *A*. See figure 40. Unscrew the barrel from the barrel extension. Remove the barrel-locking spring by sliding it forward out of its seat in the barrel extension. Push out the breech lock pin and remove the breech lock.

Q. Explain the disassembly of the cover group. *A*. See figure 41. Remove the cover. Withdraw the belt feed lever pivot stud cotter pin and pry the belt feed lever off its stud, taking care that the lever plunger and spring do not fly out. In the removal of the belt feed lever the toe of the lever must be in line with the slot in the cover to withdraw it. Remove the belt feed slide and drive out the belt feed pawl pin. This allows the belt feed pawl, spring, and pawl arm to be separated. Drive out the cover latch pin and remove the cover latch. Remove the cover latch spring by lifting its front end out of the slot in the cover and sliding it forward. Remove the cover extractor spring by releasing its rear end from its seat in the cover extractor cam and sliding it to the rear.

Q. Explain the disassembly of the backplate group. *A.* See figure 42.

(1) Drive out the backplate latch pin and remove the backplate latch and backplate latch spring, taking care that the spring does not fly out when the latch is removed. Drive out the latch lock pin and remove the latch lock, latch lock spacers, and latch lock spring. Drive out the trigger pin, being careful that the trigger spring does not fly out upon removal of the trigger.

(2) The adjusting screw, buffer disks, and buffer plate are not removed except for repair or adjustment. If it is necessary to remove these parts, unscrew the adjusting screw and remove the adjusting screw plunger and the adjusting screw plunger spring. Remove the buffer disks and the buffer plate through the rear end of the buffer tube.

(3) When the backplate group is fully assembled, the adjusting screw must always be kept screwed tight against the buffer disks. If the top of the adjusting screw becomes flush with the end of the buffer tube when screwed tight, remove the adjusting screw and add one or more buffer disks.

Q. Explain how to disassemble the casing and water jacket group. *A.* See figure 43.

(1) To remove the side-plate trigger assembly from the casing, withdraw the cotter pin from the side-plate trigger bolt and unscrew the side-plate trigger nut. To disassemble the assembly, unscrew the side-plate trigger extension screw, lift off the side-plate trigger extension, and remove the side-plate trigger slide spring. Push the side-plate trigger slide out of its guides in the housing. Swing the side-plate trigger cam upward and lift out the side-plate trigger spring. The side-plate trigger cam can be removed by driving out the side-plate trigger pin, but this is not done unless necessary for replacement of the cam.

(2) To remove the retracting slide assembly from the casing, pull out the locking wires and unscrew the retracting slide bracket screws. To disassemble the assembly, withdraw the cotter pin from the retracting slide bracket bolt, unscrew the retracting slide nut, and remove the bolt from the retracting slide bracket. Withdraw the cotter

pin from the retracting slide lever stud, unscrew the retracting slide nut, and remove the washer. Remove the retracting slide lever and the grip assembly from the lever stud. Remove the lever spring. Remove the retracting slide from the retracting slide bracket. Unscrew the retracting slide lever stop from the retracting slide.

(3) To remove the trigger bar from the casing, lift the end of the trigger bar pin lock from its seat on the side plate, rotate downward about 90°, and pull out the trigger bar pin.

(4) To remove the switch and switch spring, withdraw the cotter pin and unscrew the switch pivot nut.

(5) To remove the belt-holding pawl, draw the pawl pin out to the rear, taking care that the belt-holding pawl spring does not fly out upon removal of the pawl.

(6) To remove the link stripper and the front and rear cartridge stops, draw the belt-holding pawl pin located on the opposite side of the feedway from the belt-holding pawl to the rear.

(7) To remove the trunnion block lock and spring, withdraw the cotter pin from the lock.

(8) To remove the cover detent pawl and spring, withdraw the cotter pin from the pawl.

(9) Unscrew the front barrel bearing lock screw jam nut and front barrel bearing lock screw. The muzzle gland, the muzzle packing ring, and the muzzle barrel packing may then be removed. This is not recommended unless necessary for repair or to repack the barrel.

Q. Explain the procedure in assembling groups and replacing them in the gun. *A.* In general the groups are assembled and replaced in the gun in the reverse order to that in which they are removed and disassembled. There are certain precautions in connection with assembling which are outlined below and must be observed in order that the parts may be placed in the gun and in order that they may function properly after the gun is assembled. In most cases, if the gun can be assembled, failure to observe these points will result in failure of the bolt to go fully forward on the closing movement. Be sure that in assembling the part groups they are all properly assembled either for a left-hand or right-hand feed. The following instructions, (1) to (8), inclusive, are for assembly for left-hand feeding.

(1) The cam groove in the bolt switch must line up with the cam groove in the bolt marked *L*. If it is opposite the groove marked *R*, lift up the bolt switch high enough to be clear of the bolt switch stud and rotate it one-half turn so that the stud enters the opposite hole in the switch.

(2) When the cover is raised, the upper end of the belt feed lever is toward the left side of the cover and the belt feed pawl arm points toward the right and is on the upper side of the pawl.

(3) The front and rear cartridge stops and the link stripper are assembled on the right side of the feedway, and the belt-holding pawl on the left side of the feedway.

(4) In assembling the barrel extension, make sure that the breech lock is inserted with the bevel faces to the front and the double bevel on top.

(5) The accelerator is assembled in the oil-buffer body with the accelerator tips up and rounded surface to the front.

(6) The cocking lever is assembled in the bolt with the rounded nose on the lower end of the lever toward the rear of the bolt so that it will properly engage the rear of the slot in the firing pin extension.

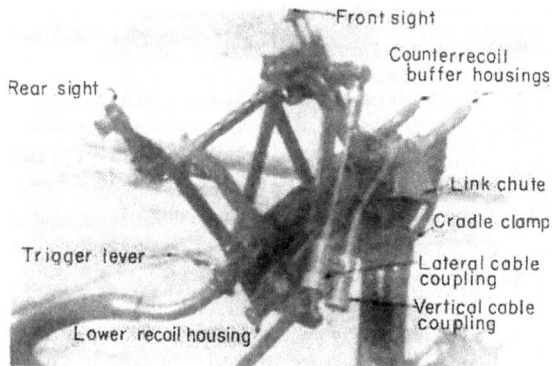

FIGURE 44.—Cradle of machine-gun Mount M2.

(7) In assembling the fiber buffer disks, be sure that they are clean and free of rough edges and surfaces. Assemble them in the tube one at a time, firmly seating each disk and using sufficient disks so that when the adjusting screw is inserted and tightened, its outer face will extend slightly outside the backplate.

(8) When reassembling the backplate to the gun, keep the latch lock in the unlocked position until the backplate is latched.

d. Machine-gun mounts.—Q. Identify the important parts of the cradle of the machine-gun mount M2. A. See figure 44.

Q. Identify the parts of the machine-gun mount M1. A. See figure 45.

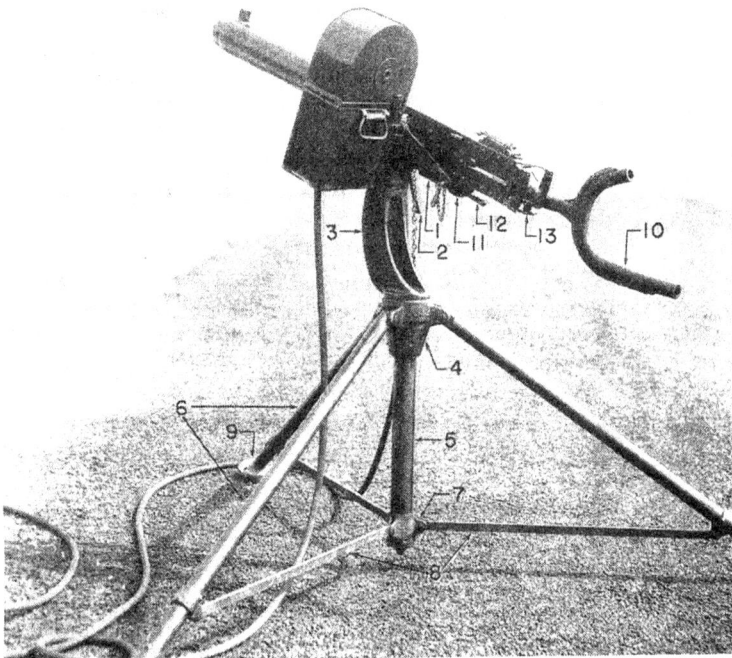

1. Cradle.
2. Yoke lock.
3. Yoke.
4. Socket.
5. Center support.
6. Tripod legs.
7. Adjusting bracket and stop collar.
8. Leg braces.
9. Shoes.
10. Shoulder stock.
11. Adjusting joint (for raising or lowering stock).
12. Adjusting joint lock.
13. Adjusting slide lock (for changing length of stock).

FIGURE 45.—Machine-gun mount M1 with shoulder stock.

Q. What is the purpose of the spring recoil mechanism incorporated in the cradle of the M2 mount? *A*. It is to reduce the vibration of the gun caused by the strong recoil and rapid rate of fire.

Q. What care is taken of the tripod or pedestal? *A*. Sand and dirt are cleaned out of the moving parts, unpainted parts are oiled well, or greased if it is to be stored.

Q. In assembling the M2 mount, what precaution is taken while assembling the legs to the pedestal? *A*. The cover is placed over the top of the pedestal to exclude dirt and grit from the cradle seat.

Q. What parts of the M2 mount should never be tampered with by members of the machine-gun squad? *A*. The recoil mechanism and the trigger control mechanism.

Q. What precautions must be taken to protect the sight mechanism of the M2 mount? *A*. When central control is used, care must be taken that the cover of the gun is not raised while the sight mechanism is being operated. At the conclusion of drill or action, if the guns and mounts are to remain in position, run the elevating screw of the front sight mechanism down as far as possible to protect it against possible injury. When movements are made from one emplacement to another, remove the sight assembly from the mount and place it in the carrying case.

21. Safety precautions.—*Q*. What is the purpose of the safety precautions discussed in this paragraph? *A*. They are prescribed for use in time of peace to insure the safety of the towing airplane and the personnel on the firing line. The fundamentals indicated should be applied under war conditions where circumstances permit.

Q. When may a gun be loaded? *A*. Only when the command to load has been given by the officer conducting the fire. Guns will always be unloaded except when firing or about to fire.

Q. What precaution should be taken to prevent live ammunition from being used inadvertently in the gun? *A*. No live ammunition should be allowed near the emplacements except when firing is to take place.

Q. When can persons go in front of the guns? *A*. No person will be allowed to go in front of the firing line until permission has been granted by an officer, who has ordered all guns to be cleared.

Note.—Members of the gun squad should always pass in rear of the gun from one side to the other.

Q. What precautions are taken with relation to the pointing of the gun? *A*.

(1) The gun is always kept pointed inside the limiting (left and right) boundaries of the field of fire except when boresighting or

other operations require it to be pointed in some other direction. Before the gun is pointed outside the limits of the field of fire, it must be carefully checked to insure that it is unloaded.

(2) A loaded gun must always be pointed at a safe part of the field of fire.

(3) The gun is depressed considerably below the elevation of the towing airplane until the airplane has passed the line of sight of the gun sights.

Q. When beginning to track a target, at what time does the gunner first place his hand on the trigger? *A.* When his sight is alined with the target (central control), or when he has taken the required estimated leads (individual tracer control), and the field of fire has been indicated as safe.

Q. What is done if the towing plane develops engine trouble? *A.* CEASE FIRING is executed immediately.

Q. When firing ceases at the firing point what precautions are taken immediately? *A.* Machine guns are unloaded, covers lifted, and bolts pulled back.

Q. Before removing the machine gun from the tripod or pedestal, what must be done? *A.* The gun must be unloaded, the cover lifted, and the chamber examined to see that it is empty.

Q. What precautions must be taken to protect against a dragging towline? *A.* When targets are towed over or close to gun positions, overhead cover of some sort must be available so that personnel will be protected from a dragging towline.

Q. If malfunctioning of the gun causes it to fire continuously while the trigger is released, how can the firing be halted? *A.* Grasp the ammunition belt near the point where it enters the feedway and twist it with force so that the rounds cannot enter the feedway.

NOTE.—As soon as fire is halted, raise the cover, remove the ammunition belt and any ammunition in the gun, and determine and correct the trouble.

CHAPTER 4

AMMUNITION

SECTION I

AMMUNITION, FUZES, AND PROJECTILES FOR GUNS

22. Ammunition, general.—*a. 3-inch gun.*—*Q.* What kind of ammunition is used for 3-inch guns? *A.* Fixed ammunition: time fuzed, high explosive shell.

Q. What is fixed ammunition? *A.* That in which the different parts of one round are assembled into one unit. The unit includes a cartridge case, with its primer and powder (propelling charge), and a projectile with its time fuze.

Q. What type of primer is used? *A.* Percussion primer.

Q. Of what does the propelling charge consist? *A.* Smokeless powder.

Q. Of what does the bursting charge of a high explosive shell consist? *A.* High explosive powder.

Q. Of what does the bursting charge of shrapnel consist? *A.* Black powder.

Q. What is the purpose of the booster in a high explosive shell? *A.* To detonate the bursting charge.

Q. What is the approximate weight of a round of 3-inch ammunition? *A.* From about 25 to about 27 pounds depending on the kind of projectile and the model of the gun.

Q. What is the normal muzzle velocity of the guns in your battery when using high explosive shell? *A.* ———.

Q. Name and point out the principal parts of a complete round of ammunition. *A.* See figure 46.

b. 105-mm gun.—*Q.* What two kinds of ammunition are furnished for the 105-mm gun? *A.* Practice shell and high explosive shell.

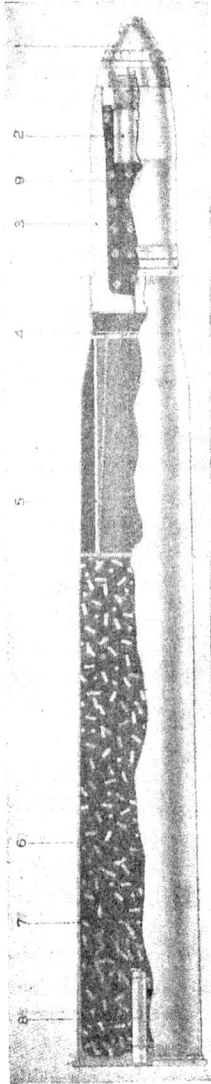

1. Time fuze Mk. III.
2. Booster Mk. X.
3. Common steel shell Mk. IX.
4. Diaphragm.
5. Distance wad.
6. Propelling charge.
7. Cartridge case.
8. Percussion primer.
9. Bursting charge.

FIGURE 46.—Complete round 3-inch ammunition.

Q. What kind of primer is used? *A.* Percussion primer.

Q. Of what does the propelling charge consist? *A.* Smokeless powder.

Q. Of what does the bursting charge of the high explosive shell consist? *A.* High explosive powder.

Q. Of what does the bursting charge of the practice shell consist? *A.* Black powder.

Q. What type of ammunition is used for target practice? *A.* The practice shell. This shell is identical with the high explosive shell M38 except that it contains a practice loading instead of a high explosive filler.

23. Fuzes.—*Q.* What type of fuze is used on gun ammunition? *A.* Time fuze.

Q. What is a time fuze? *A.* A fuze placed on the nose of the projectile, which can be set to explode the projectile at a desired fuze range or a certain number of seconds after the gun is fired.

Q. What two types of time fuzes are employed with 3-inch guns? *A.* Powder-train fuze and mechanical fuze.

Q. What type of time fuze is employed with 105-mm guns? *A.* Mechanical fuze.

Q. Are fuzes assembled to the projectile for shipment? *A.* Yes.

Q. Describe the Mk. III fuze. *A.* The Mk. III fuze is a powder-train fuze having two train-powder rings, (6) and (8) (fig. 47), and a magazine charge (10). When the fuze is set at *safe* the lower powder train (7) in the ring (8) is completely cut off from the lower ignition pellets. The upper ring could burn completely without igniting the charge. Turning the graduated ring (8) sets the fuze, determining the time the trains (7) must burn before reaching the ignition pellets (13) and (14) and igniting the charge.

Q. What starts the fuze burning? *A.* The set-back or shock of firing the shell out of the gun causes the plunger (2) to strike the firing pin (5), which sets off the primer (4) and the ignition powder pellet (12), and starts the powder trains burning.

Q. What is the fundamental factor of operation of the mechanical time fuze? *A.* The mechanism is similar to that of a clock and can be set to cause the shell to explode at a predetermined time.

Q. What is the source of power for the timing element in the mechanical fuze? *A.* The rotation of the projectile powers the timing element.

Q. What are some of the advantages of the mechanical fuze over the powder-train fuze? *A.* The mechanical fuze has greater accuracy of timing; atmospheric conditions do not affect the timing mechanism; it can be stored for long periods without deterioration; and the body and lower cap are fitted with slots instead of lugs which eliminates the difficulty of sheared lugs often experienced with powder-train fuzes. Also the timing element operates to a maximum of 30 seconds as compared with 21 seconds for the powder-train fuze.

24. Precautions in handling.—*Q.* When is the moisture resistant seal of the fiber ammunition containers broken? *A.* Not until it is necessary to prepare the ammunition for use.

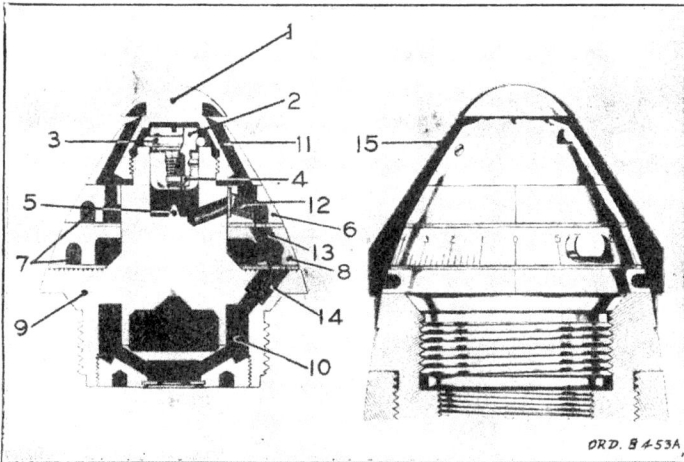

1. Closing cap.	9. Body.
2. Concussion plunger.	10. Magazine charge (black powder).
3. Resistance ring.	11. Vents.
4. Concussion primer.	12. Powder pellet.
5. Concussion firing pin.	13. Powder pellet.
6. Upper time train ring.	14. Powder pellet.
7. Powder train.	15. Waterproof cover.
8. Lower or graduated time train ring.	

FIGURE 47.—Time fuze Mk. III.

Q. What precaution should be taken with reference to high temperatures? *A.* Protect ammunition, particularly fuzes, from high temperatures, including direct rays of the sun.

Q. Are fuzes ever disassembled by coast artillery personnel? *A.* No.

Q. What precaution concerning the cleanliness of ammunition must be taken? *A.* The ammunition must be free of sand, mud, grease, or other foreign matter when loaded into the gun.

Q. What is done with unfired ammunition at the end of a day's firing? *A.* Rounds prepared for firing but not fired are returned to

their original packings and appropriately marked. These rounds
are used first in subsequent firings, in order that stocks of opened
packings may be kept to a minimum.

Q. What keeps the plunger of the fuze from setting the fuze off if
the shell should be dropped or otherwise receive a blow on the nose?
A. The resistance ring (3)(fig. 47) holds the plunger up unless the
fuze gets a very heavy shock, like that in firing. If the rings are
set at *safe*, however, even then no flame could get through to the
lower ignition powder pellet.

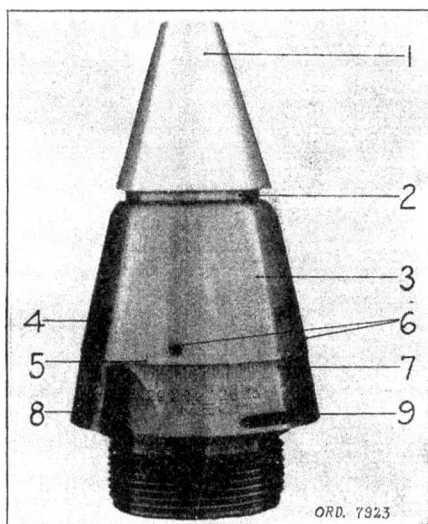

1. Upper cap.	6. Setscrews.
2. Groove for fuze setter locking lever.	7. Body.
3. Lower cap.	8. Slot for fuze setter lug.
4. Slot for fuze setter lug.	9. Fuze wrench slot.
5. Register line.	

FIGURE 48.—Mechanical time fuze M43.

Q. Is it possible for the fuze to start burning without the gun being
fired? *A.* Yes, a hard blow might cause the fuze to start burning.

Q. What should be done in such a case? *A.* The projectile should
be inserted in the gun and fired.

Q. What should be done if a fuze has been cut and the round inserted in the gun and CEASE FIRING is given? *A.* The round should
be removed from the gun and set at *safe*.

Q. How should ammunition be handled? *A.* With care always,
particular care being taken not to damage the fuzes, or to strike
the primers.

SECTION II

AMMUNITION. CHARACTERISTICS AND SERVICE, AUTOMATIC WEAPONS

	Paragraph
General	25
Characteristics	26
Handling and inspection of ammunition	27

25. General.—*Q.* What types of ammunition are normally used in automatic weapons? *A.*

(1) *Machine guns, caliber, .30 and caliber .50.*
    Ball.
    Tracer.

(2) *37-mm guns.*
    Service ammunition (all tracer).
    Target practice ammunition (all tracer).

Q. What other type of ammunition might be used? *A.* Armor-piercing.

Q. When would armor-piercing projectiles be used? *A.* Against certain types of airplanes which are armored. Also against tanks and similar targets.

Q. What is the purpose of marking ammunition for target practice with colored printer's ink? *A.* If bullets are dipped in specially prepared printer's inks of different colors, and each gunner fires bullets painted a distinctive color, the hits obtained on a sleeve target by each of the gunners of the fire unit can be readily determined.

Q. What colors are suitable for use? *A.* Red, orange, green, and blue. Unpainted bullets may be substituted for one of the colors when desired.

26. Characteristics.—*Q.* What explosive is provided in the projectiles of automatic weapons? *A.*

(1) *37-mm gun.*—High explosive.
(2) *Caliber .50 machine gun.*—None. Solid bullet.
(3) *Caliber .30 machine gun.*—None. Solid bullet.

Q. What are the weights of the rounds of the various automatic weapons? *A.*

(1) *37-mm gun.*—2.62 pounds.
(2) *Caliber .50 machine gun.*—Approximately 4 ounces.
(3) *Caliber .30 machine gun.*—Approximately 1 ounce.

Q. What are the approximate maximum tracer ranges of automatic weapon tracer shell or bullets? *A.*

(1) *37-mm gun.*—3,500 yards.
(2) *Caliber .50 machine gun.*—1,850 yards.

(3) *Caliber .30 machine gun.*—1,000 yards.

Q. What types of clips or belts are employed with the various automatic weapons? *A*.

(1) *37-mm gun.*—Ammunition clips holding 10 rounds each.

(2) *Caliber .50 machine gun.*—Disintegrating link belts holding 200 rounds each. The number of rounds is limited by the size of the ammunition box.

(3) *Caliber .30 machine gun.*—Fabric belts holding 250 rounds each.

Q. What type of fuze is provided for ammunition of automatic weapons? *A*.

(1) *37-mm gun.*—Supersensitive fuze. These fuzes are so sensitive that they will function upon striking an object even as light as the fabric wing of a plane or the envelope of a balloon.

(2) *Caliber .50 machine gun.*—None.

(3) *Caliber .30 machine gun.*—None.

Q. What is the purpose of the self-destroying feature of 37-mm projectiles? *A*. To eliminate the danger to friendly ground elements from rounds which do not strike the hostile air target. This self-destroying element functions when the tracer burns out, at a range of about 3,500 yards.

27. Handling and inspection of ammunition.—*Q*. How is the ammunition for automatic weapons packed? *A*.

(1) *37-mm.*—20 in a metal-lined wooden box weighing 85 pounds.

(2) *Caliber .50.*—280 in a metal-lined wooden box weighing 106 pounds.

(3) *Caliber .30.*—1,200 in a metal-lined wooden box weighing 100 pounds.

Q. What equipment is provided for loading machine gun and 37-mm ammunition into clips or belts? *A*. Belt-filling machines are provided for loading caliber .30 fabric belts. Link-loading machines are provided for loading caliber .50 link belts. 37-mm ammunition must be loaded into the ammunition clips by hand.

Q. When inspecting ammunition prior to firing, what faults should be looked for? *A*. Loose rounds, short rounds, thick or thin rims, inset primers, split cases, and battered or dented cases in machine gun ammunition. If caliber .50 ammunition has been removed from link belts and then reloaded, a number of short rounds may be expected. The most frequent fault in 37-mm ammunition is inset primers.

Q. State some of the precautions to be taken in handling ammunition. *A*.

(1) Do not open the sealed containers until ready to use the ammunition.

(2) Protect the ammunition from mud, sand, dirt, and water, carefully wiping off the cartridges if they get wet or dirty.

(3) Do not try to use rounds with dented cases or those in which the projectiles are loose.

(4) Use no grease, oil, or other lubricant on the cartridges.

(5) Do not expose ammunition to direct sunlight or any other heat for any length of time.

SECTION III

TRANSPORTING, HANDLING, AND STORING AMMUNITION

28. Transporting ammunition.—*Q.* What regulations govern the handling of ammunition? *A.* TM 4–205, TM 9–2900, AR 30–955, AR 30–1270, AR 700–10, and such local regulations as may be prescribed; for example, many localities require a special placard or flag to be displayed on a vehicle transporting ammunition.

Q. Where may detailed regulations prescribing the transportation of explosives be obtained? *A.* From the Interstate Commerce Commission through The Quartermaster General or the Chief of Ordnance. These regulations permit the Government to prescribe its own shipping regulations, marking, packing, and storing, but the War Department regulations comply in general with the Interstate Commerce Commission regulations.

Q. How are the necessary labels obtained? *A.* On requisition through The Quartermaster General.

Q. What responsibilities must the shipping officer assume? *A.* That all regulations are complied with. In case of fire or accident the shipping officer is responsible.

Q. May explosives be carried as a deckload on Army transports? *A.* No.

Q. When transporting explosives by truck what procedure will be followed? *A.*

(1) Comply with local regulations and all Army Regulations.

(2) Contact local authorities and select safe routes.

(3) Take every precaution against fire.

Q. What precautions must be taken against fire? *A.*

(1) Inspect trucks daily for wiring, lights, brakes, gasoline tanks, and lines.

(2) Keep vehicle and engine clean.

(3) Permit no smoking.

(4) Keep safety matches in a metal container in tool box.

(5) Provide each truck with a sand box (3 cubic feet) and a shovel.

(6) Instruct all drivers in fire fighting. Ammunition requires considerable heat before it will explode, and a fire, if discovered in time, can usually be put out with safety.

(7) Do not transport detonating agents with other explosives.

(8) Lay boards over all iron parts of the truck.

(9) See that load is well braced and stayed and is covered with a tarpaulin to prevent fire by sparks.

Q. When ammunition is being transported by convoy what precautions should be taken? *A.*

(1) Keep a safe distance between trucks to avoid danger of collision.

(2) Stop once each hour and inspect the load.

(3) Do not stop in populous areas.

(4) Permit no unauthorized riders.

(5) If a truck breaks down transfer its load to another truck. Do not attempt to tow.

(6) In case of fire all other vehicles will proceed to a safe distance and guards will be posted at a safe distance from the fire to ward off other traffic.

Q. How is artillery ammunition packed in a truck for transportation? *A.* Laid on its side parallel to the sides of the truck. If more than one layer is to be placed in the truck, strips of planking should be laid to protect the rotating bands.

29. Handling ammunition.—*Q.* Under whose supervision should ammunition be handled? *A.* Under a competent person who understands thoroughly the hazards and risks involved.

Q. Name some hazardous explosives. *A.* Detonators, bulk explosives, and smokeless powder.

Q. What precautions should be observed by personnel engaged in handling explosives? *A.* No metal tools of any kind should be used by any personnel engaged in handling explosives. Extreme care should be taken to insure that such personnel do not have on or about their persons any metal tools, nails, matches, cartridges, firearms, or similar material and that their shoes are not shod with iron nails or other metallic substances which are liable to cause a spark. Only shoes which have soles of felt or soft leather should be worn.

Q. In case explosives are spilled from a container, what should be done? *A.* All work must be stopped until the explosives have been swept up and the area has been neutralized.

Q. Where may damaged containers be repaired? *A.* In the open, or in a building especially provided for this purpose, at least 100 feet from the magazine. boat. or truck containing ammunition.

30. Storing ammunition.—*Q.* If ammunition must be stored outside. is it necessary to protect it from the sun? *A.* Yes. It must be protected from the direct rays of the sun in such a way that air may freely circulate through the pile.

Q. What ammunition may be stored in a battery storeroom? *A.* The small quantity of small-arms ammunition required for current use.

Q. Where must ammunition be stored? *A.* In special magazines such as described in Technical Manuals.

Q. How is ammunition segregated in storage? *A.* Ammunition is placed in neat, stable piles by a lot number and is raised off the floor on 2-inch battens.

Q. How high may ammunition be piled? *A.* This depends on the strength of the container, but piles should not exceed the height of the eaves in magazines.

Q. What does an acid odor in a powder magazine indicate? *A.* Danger—powder is decomposing.

Q. What testing instruments are placed in powder and ammunition magazines? *A.*

(1) Maximum and minimum thermometer.

(2) Hygrometer.

(3) Litmus paper.

Q. How is air circulation provided in ammunition storage? *A.* By dunnage or by cleats on the boxes.

Q. Where may small-arms ammunition be stored? *A.* In any magazine or warehouse which offers good protection against the weather.

Q. In case of doubt of the condition of ammunition in storage, who is notified? *A.* The local ordnance officer.

CHAPTER 5

GUNNERY, FIRE CONTROL, AND POSITION FINDING—
GUNS

SECTION I

DUTIES OF RANGE SECTION

31. Duties of range section.—Instruction in duties of the range
section will be three-quarters practical and one-quarter theoretical and
will be limited to the instruments with which the battery is equipped.
Each candidate should be required to level each of the instruments,
read either dials or scales, connect up the data transmission systems,
and operate the power plant.

Q. What are the posts of the range section? *A.* See figures 49 to
53, inclusive.

Q. What are the duties of the range section at the various commands? *A.* See tables VIII, IX, X, XI or XII.

NOTE.—Drills in tables VIII, IX, and X pertain specifically to certain equipment in the range section. Other men are available in the range section who are
not specifically assigned to one piece of equipment. In handling the equipment
it should be kept in mind that the minimum number of men to handle either the
director, height finder tube, or power plant is 8. To endeavor to maneuver the
equipment with less men is to invite injury to the men and to the equipment.

TABLE VIII.—*Drill for range section, 3-inch gun battery with director M4*

Details	PREPARE FOR ACTION	DETAILS, POSTS	TARGET	CEASE TRACKING	MARCH ORDER
Chief of range section.	Repeats the command. Supervises work of entire range section. Indicates positions for various instruments. Gives command to turn on power. When section is in order, he reports to range officer, "Sir, range section in order."	Repeats the command. Supervises work of section.	Repeats the command. Assists in getting director on target. Reports to range officer, "Sir, range section on target" as soon as both director and height finder locate target and commence tracking. In addition, may be required to assist in fire adjustment.	Repeats the command.	Repeats the command. Supervises replacing equipment.
Instrument corporal, No. 1 (in charge of director).	Supervises setting up director. Levels director. Using data furnished him, orients director. Assisted by height finder sergeant and gun commanders, checks synchronization of entire electrical data transmission system. Reports to chief of range section, "Director in order."	Repeats the command. Supervises work of manning detail of director.	Repeats the command. Turns on power at director. Gives such directions as necessary to assist in getting director on target. Reports to chief of range section, "On target" as soon as trackers commence to track target. In addition, may be required to assist in fire adjustment and spotting.	Repeats the command. Turns power off at director about 10 seconds after tracking has stopped.	Repeats the command. Supervises replacing director.
Elevation tracker, director.	Assists in placing director and power plant. Assists instrument corporal in orienting and synchronizing director.	Takes post on left side of director facing elevation tracking telescope.	Looks through telescope and elevates or depresses as directed. When he sees the target, calls out "On target." Thereafter tracks target bisecting it with horizontal cross hair. Uses "slow motion" whenever possible in order to target track smoothly.	Ceases tracking. Remains at post.	Assists in replacing director and power plant.
Azimuth tracker, director.	Assists in placing director. Assists in removing height finder tube from carrying case and placing it on cradle. Assists in placing power plant in position.	Takes post on right side of director facing azimuth tracking telescope.	Looks through telescope and traverses as directed. When he sees target, calls out "On target." Thereafter tracks target bisecting it with the vertical cross hair. Uses "slow motion" whenever possible in order to track target smoothly.	-----do------	Assists in replacing director, height finder tube, and power plant.

113

TABLE VIII.— *Drill for range section, 8-inch gun battery with director M4*—Continued

Details	PREPARE FOR ACTION	DETAILS, POSTS	TARGET	CEASE TRACKING	MARCH ORDER
Range setter, director.	Assists in placing director. Assists in removing height finder tube from carrying case and placing it on cradle. Assists in placing power plant.	Takes post in front of director facing angular height dials.	As soon as director is on target, matches angular height dials (both coarse and fine) using range setting handwheel. Thereafter, by regulating range rate knob and handwheel, keeps angular height dials matched continuously. Uses prediction button (red button) as required.	Sets range rate knob to zero. Ceases to match angular height dials. Remains at post.	Assists in replacing director, height finder tube, and power plant.
Altitude setter, director.	Assists in placing director. Assists in removing height finder tube from carrying case and placing it on cradle. Assists in placing power plant in position.	Takes post in front of director facing altitude dial.	As soon as altitudes are transmitted from height finder, matches pointers of altitude dial using altitude setting handwheel.	Ceases to match pointer of altitude dial. Remains at post.	Do.

TABLE IX.— *Drill for height finder M1 or M2*

Details	PREPARE FOR ACTION	DETAILS, POSTS	TARGET	CEASE TRACKING	MARCH ORDER
Sergeant, height finder	Supervises setting up, orienting, and leveling height finder. Assists instrument corporals in checking synchronization of all electrical data transmission between height finder and director. After observer has made his individual adjustments, reports to chief of range section, "Height finder in order."	Repeats the command. Supervises work of height finder detail.	Repeats the command. Assists trackers in getting on target. As soon as height finder is on target, he reports to chief of range section, "Height finder on target."	Repeats the command.	Repeats the command. Supervises replacing height finder.
Observers, stereoscopic (2) (1 as relief).	Assists in setting up height finder. Levels instrument. Using data furnished him, orients height finder. Makes all adjustments necessary for operation of instrument.	Takes post facing height finder at observer's eyepiece.	When trackers report, "On target," looks through eyepiece and establishes and maintains stereoscopic contact with target. Indicates to height setter when to send data to director.	Ceases tracking. Remains at post.	Assists in replacing height finder.

Elevation tracker, height finder.	Assisted by azimuth tracker, height finder, sets up tripod and cradle for height finder. Assists in removing height finder tube from carrying case and placing it on cradle. Procures his telescope from packing chest and mounts it on height finder.	Takes post in rear of height finder facing elevation tracker's telescope.	If so directed by range officer, performs stereoscopic spotting in addition to altitude determination. If director is on target, matches pointers of elevation receiver of target designating system. If not, searches in designated area for target. Once target is located, tracks it accurately and smoothly in elevation.	....do............	Removes telescope and replaces it in packing chest. Assists in replacing height finder tube in carrying case. Assisted by azimuth tracker, dismantles cradle and tripod and replaces them in packing cases.
Azimuth tracker, height finder.	Assists elevation tracker, height finder, in setting up tripod and cradle for height finder. Assists in removing height finder tube from carrying case and placing on cradle. Procures his telescope from packing chest and mounts it on height finder.	Takes post in rear of height finder facing azimuth tracker's telescope.	If director is on target, matches pointers of azimuth receiver of target designating system. If not, searches in designated area for target. Once target is located, tracks it accurately and smoothly in azimuth.	....do............	Removes telescope and replaces it in packing chest. Assists in replacing height finder tube in carrying case. Assists elevation tracker to dismantle cradle and tripod and replace them in packing cases.
Height setter, height finder.	Assists in removing height finder tube from carrying case and placing it on cradle. Connects plug for wiring circuit from cradle to height finder tube. Lays cable from height finder to director and connects cable to height finder and director.	Takes post in front of height finder facing altitude transmitter.	Assists in locating target and getting height finder on target. Sets altitude transmitter to values indicated on altitude scale when directed by observer.	Ceases to send altitudes to director. Remains at post.	Removes cable from height finder. Disconnects plug for wiring circuit from cradle to height finder tube. Assists in replacing height finder tube in carrying case. Disconnects cable from height finder at director and replaces cable on reel.

Table X.—*Drill for members of range section using M4 director, not included in director or height finder details*

Details	PREPARE FOR ACTION	DETAILS, POSTS	TARGET	CEASE TRACKING	MARCH ORDER
Instrument corporal No. 2 (in charge of power plant and cables).	Supervises laying cables and connecting receptacles. Supervises setting up and operation of power plant.	Takes post at power plant.	Remains at power plant.	No duties.	Supervises removal of receptacles and picking up cables. Supervises replacing cable reels and power plant.
Cable private (assistant to IC No. 2).	Procures and places main junction box. Assists instrument corporal No. 2 in laying cables and connecting receptacles.	Takes post in vicinity of director.	Insures proper power is being delivered to director.	do	Assists instrument corporal No. 2 in removing receptacles and picking up cables.
Power plant operator.	Assists in placing power plant. With help of assistant power plant operator, lays cable from power plant to main junction box. Connects receptacles to power plant and main junction box. When directed by chief of range section, starts power plant.	Takes post at power plant.	Operates power plant as directed	do	Stops power plant. Removes receptacles of cables and with help of his assistant reels up cable. Assists in replacing power plant.
Assistant power plant operator.	Assists in placing director. Assists in removing height finder tube from carrying case and placing it on cradle. Assists in placing power plant.	do	Assists power plant operator in his duties.	do	Assists in replacing director, height finder tube, and power plant.
Instrument corporal No. 3 (from headquarters section).	Sets up, levels, and orients BC telescope. Reports to chief of range section. "O_1 station in order."	Takes post at BC telescope.	Reports to chief of range section, "O_1 station on target." Reports observed deviations.	Ceases to observe. Remains at post.	Dismounts instrument.
BC spotter and observer.	Assists instrument corporal at BC telescope.	do	Tracks target	Ceases tracking. Remains at post.	Assists in dismounting BC telescope.
Flank spotters and observers (6) (see note 1).	Set up, level, and orient instruments. Report to chief of range section, "..... station in order."	Take posts at instruments.	"Track target. Report to chief of range section, "..... station on target." Report continuously the values of ϵ_m and θ. Report observed deviations.	do	Dismount instruments.

Details	PREPARE FOR ACTION	DETAILS, POSTS	TARGET	CEASE TRACKING	MARCH ORDER
Chauffeurs (4)	Assist in placing director. Assist in removing height finder tube from carrying case and placing it on cradle. Assist in placing power plant in position. Perform any other duties as directed by chief of section.	Proceed with trucks to designated covered positions. Personnel returns to battery position.	NOTE.—Replacements for various trackers or setters at director or height finder as needed. May assist in spotting for fire adjustment.	No duties.	Bring trucks to battery position. Assist in replacing director, height finder tube, and power plant. Perform any other duties as directed by chief of range section.

This table is merely a guide as to the disposition of men to be used. Local conditions may make it advisable to use the personnel of the range section in a different manner to take advantage of special qualifications of certain men.

NOTES

1. Includes three basics to assist the flank spotters and observers.
2. The range section also includes a staff sergeant (electrician) in charge of maintenance of the electrical equipment.

TABLE XI.— *Drill for director M3 (3-inch guns) and M3A1 (105-mm guns)*

Details	PREPARE FOR ACTION	DETAILS, POSTS	TARGET	CEASE TRACKING	MARCH ORDER
Chief of range section.	Repeats the command. Supervises work of entire range section. Indicates positions for various instruments. Gives command to turn on power. When section is in order, he reports to range officer, "Sir, range section in order."	Repeats the command. Supervises work of section.	Repeats the command. Assists in getting director on target. Reports to range officer, "Sir, range section on target," as soon as both director and height finder locate target and commence tracking. In addition, may be required to assist in fire adjustment.	Repeats the command.	Repeats the command. Supervises replacing of equipment.
Instrument corporal (in charge of director) No. 1.	Supervises setting up of director. Levels director. Using data furnished him, orients director. After power has been turned on, assisted by height finder observer and gun commanders, checks synchronization of entire electrical data transmission system. Reports to chief of range section, "Director in order."	Repeats the command. Supervises work of manning detail of director. Takes place at rate setter handwheels to take the place of Nos. 18 and 19 until they return from parking their trucks.	Repeats the command. Turns on power at director. Gives such instructions as necessary to assist in getting director on target. Reports to chief of range section, "On target," as soon as trackers commence to track target. In addition, may be required to assist in fire adjustment and spotting. Until the return of Nos. 18 and 19, matches follow pointers with tachometer pointers on the rate setter dials. After Nos. 18 and 19 take over dials, returns to supervision of director.	Repeats the command. Turns off power at director about 10 seconds after tracking has stopped.	Repeats the command. Supervises replacing of director.

TABLE XI.—*Drill for director M8 (8-inch guns) and M8A1 (105-mm guns)*—Continued

Details	PREPARE FOR ACTION	DETAILS, POSTS	TARGET	CEASE TRACKING	MARCH ORDER
Elevation tracker, director, No. 4.	Assists in placing director. Assists instrument corporal No. 1 in orienting and synchronizing director.	Takes post on left side of director, facing elevation tracking telescope.	Looks through telescope and elevates or depresses as directed. When he sees target, calls out, "On target." Thereafter, tracks target, bisecting it with horizontal cross hair. Uses "slow motion" whenever possible in order to track target smoothly.	Ceases tracking. Remains at post.	Assists in replacing director and power plant.
Azimuth tracker, director, No. 5.	Assists in placing director. Assists in removing height finder tube from carrying case and placing it on cradle. Assists in placing power plant in position.	Takes post on right side of director facing azimuth tracking telescope.	Looks through telescope and traverses as directed. When he sees target, calls out "On target." Thereafter, tracks target, bisecting it with the vertical cross hair. Uses "slow motion" whenever possible in order to track target smoothly.	....do......	Assists in replacing director, height finder tube, and power plant.
Altitude setter, director, No. 6.	Assists in placing director. Assists in removing height finder tube from its carrying case and placing it on cradle. Assists in placing power plant in position.	Takes post on left side of director facing altitude dial.	Matches the outside (mechanical) pointer on the altitude dial with the inside (electrical) pointer. Whenever the altitude changes decidedly, due either to a correction being applied or to the actual maneuvers of the target, he reports, "Altitude change," to the rate setters.	Ceases matching pointers. Remains at post.	Do.
Angular height setter, director, No. 7.	Assists in placing director. Assists in removing height finder tube from its carrying case and placing it upon the cradle. Assists in placing the power plant in position.	Takes post on left side of director facing the present angular height dial.	Turns on the range-rate motor switch. Using the present horizontal range handwheel, he matches the pointers of the angular height dials with the outer pointers. Once matched he uses the range-rate handwheel to keep them matched. When he has established a uniform rate, he reports, "Angular height O. K." and maintains as steady a change of rate as possible.	Ceases matching pointers. Turns off range-rate motor. Remains at post.	Do.

Rate setter, director (post taken by chauffeur No. 18). (In absence of chauffeurs Nos. 18 and 19, the posts are taken by instrument corporal No. 1 as it is possible for one man to operate both rate dials.)	Assists in placing director. Assists in removing height finder tube from its carrying case and placing it on the cradle. Assists in placing power plant in position.	Proceeds with truck to designated covered position. Returns to director and takes position on right side of director facing N-S rate dial.	Turns on the follow-up motor switch and pushes in the damper switch. As soon as No. 7 reports, "Angular height O. K.," matches the follow pointers with the tachometer pointers turning the rate handwheel so as to make the follow pointer go around the tachometer dial in the same direction as the tachometer pointer. After the rates have been once matched, pulls out the damper switch. Turns the handwheel slowly and steadily to follow the general trend of the readings disregarding slight variations between successive readings and erratic readings caused by large changes in altitude reported by No. 6.	Ceases matching pointers. Remains at post.	Brings truck to battery position. Assists in replacing director, height finder tube, and power plant. Performs any other duty as directed by chief of range section.
Rate setter, director (post taken by chauffeur No. 19).	Assists in placing director. Assists in removing height finder tube from its carrying case and placing it on the cradle. Assists in placing power plant in position.	Proceeds with truck to designated covered position. Returns to director and takes position on right side of director facing E-W rate dial.	As soon as No. 7 reports, "Angular height O. K.," presses the tachometer lever fully down and quickly releases it. Matches the follow pointer with the tachometer pointer, turning the rate handwheel so as to make the follow pointer go around the tachometer dial in the same direction as the tachometer pointer. Presses the tachometer lever again, repeating the operation as long as tracking continues. Turns handwheel slowly to follow the general trend of the readings disregarding slight variations between successive readings and erratic readings caused by large changes in altitude reported by No. 6.	........do........	Do.

TABLE XII.—*Drill for members of range section equipped with M8 director, not included in director or height finder details*

[Drill for height finder detail is the same regardless of type director used. For reference see table IX]

Details	PREPARE FOR ACTION	DETAILS, POSTS	TARGET	CEASE TRACKING	MARCH ORDER
Instrument corporal No. 2 (in charge of power plant and cables).	Supervises laying of cables and connecting receptacles. Supervises setting up and operation of power plant.	Takes post at power plant.	Remains at power plant.	No duties.	Supervises removal of receptacles and picking up cables. Supervises replacing cable reels and power plant.
Cable private.	Procures and places main junction box. Assists instrument corporal No. 2 in laying cables and connecting receptacles.	Takes post in vicinity of director.	Insures proper power is being delivered to the director.	do	Assists instrument corporal No. 2 in removing receptacles and picking up cables.
Power plant operator.	Assists in placing power plant. With help of assistant power plant operator, lays cable from power plant to main junction box. Connects receptacles to power plant and main junction box. When directed by chief of range section, starts power plant.	Takes post at power plant.	Operates power plant as directed.	do	Stops power plant. Removes receptacles of cable and with the aid of his assistant, reels up cable. Assists in replacing power plant.
Assistant power plant operator.	Assists in placing director. Assists in removing height finder tube from its carrying case and placing it on cradle. Assists in placing power plant.	do	Assists power plant operator in his duties.	do	Assists in replacing director, height finder tube, and power plant.
Instrument corporal No. 3 (from headquarters section).	Sets up, levels, and orients the BC telescope. Reports to chief of range section, "O_1 station in order."	Takes post at BC telescope.	Reports to chief of range section, "O_1 station on target." Reports observed deviations.	Ceases to observe. Remains at post.	Dismounts instrument.
BC spotter and observer.	Assists instrument corporal at BC telescope.	do	Tracks target.	Ceases tracking. Remains at post.	Assists in dismounting instrument.
Flank spotters and observers (6) (see note 1).	Set up, level, and orient instruments. Report to chief or range section, "_____ station in order."	Take posts at instruments.	Track target. Report to chief of range section, "_____ station on target." Report continuously e_m and θ. Report observed deviations.	do	Dismount instruments.

Chauffeurs Nos. 18 and 19 (serve on director detail as rate setters during firing).	Assist in placing director. Assist in removing height finder tube from its carrying case and placing it on the cradle. Assist in placing power plant in position.	Proceed with trucks to designated covered position. Return to director and take up positions as rate setters.	Perform rate setters duties as described for director detail.	Cease matching pointers. Remain at posts.	Bring trucks to battery position. Assist in replacing director, height finder tube, and power plant. Assist or perform any other duty as directed by the chief of range section.
Chauffeurs, Nos. 20 and 21	Assist in placing director. Assist in removing height finder tube from its carrying case and place it on the cradle. Assist in placing power plant in position. Perform any other duties as directed by chief of range section.	Proceed with trucks to designated covered positions. Personnel returns to battery position	NOTE.—Replacements for various trackers or setters at director or height finder as needed. May assist in spotting for fire adjustment.	No duties	Bring trucks to battery position. Assist in replacing director, height finder tube, and power plant. Perform any other duties as directed by chief of range section.

NOTES

1. Includes 3 basics to assist the flank spotters and observers.
2. The range section includes also a staff sergeant (electrician) in charge of maintenance of the electrical equipment.

121

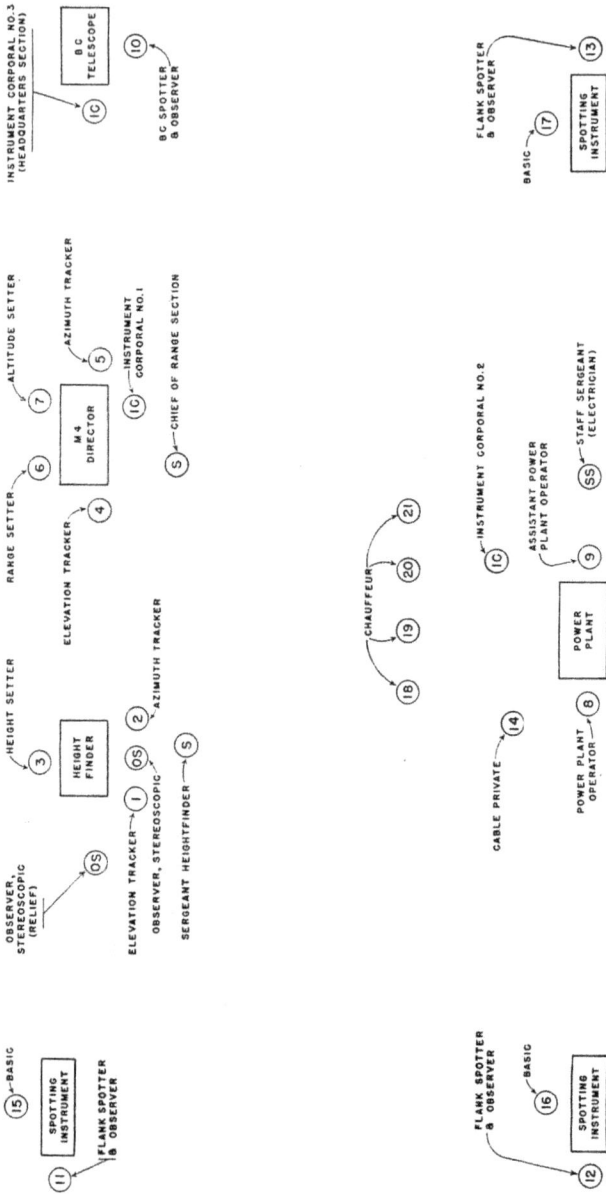

FIGURE 52.—Posts of range section equipped with M4 director.

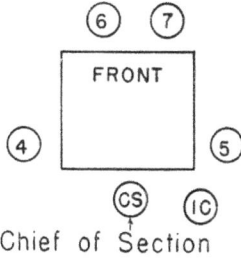

FIGURE 49.—Posts for manning detail, director M4.

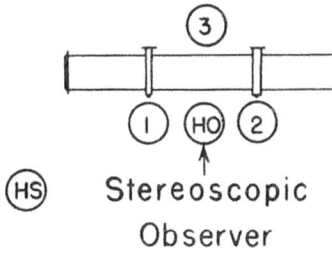

FIGURE 50.—Posts for manning detail, stereoscopic height finder.

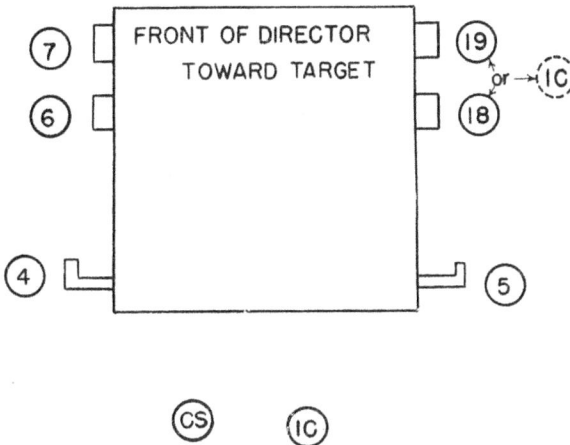

FIGURE 51.—Posts for manning detail, director M3.

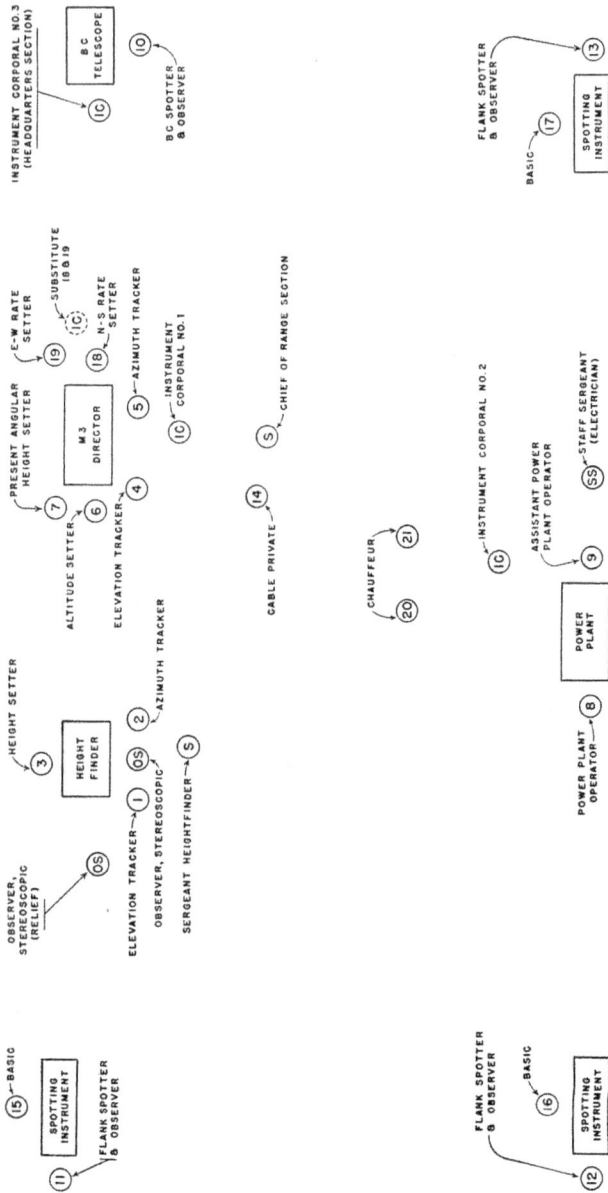

FIGURE 53.—Posts of range section equipped with M3 director.

NOT TO SCALE

Section II

POSITION-FINDING APPARATUS

32. Director M4.—*Q*. What data are computed by the director for transmission to the guns? *A*. Angle of train (A_t), quadrant elevation (ϕ), and fuze range (F).

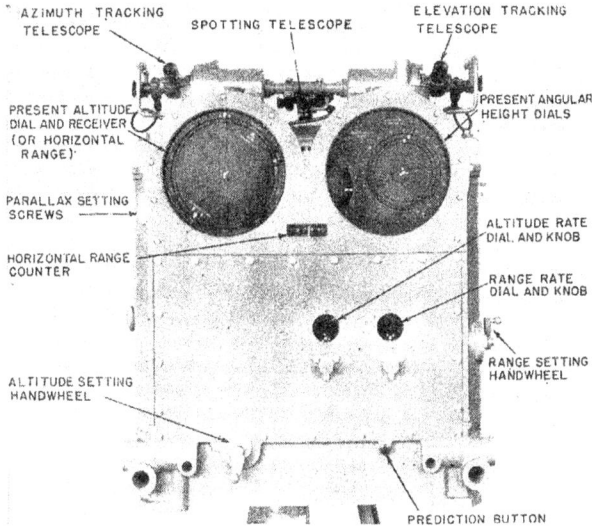

FIGURE 54.—M4 director (front).

Q. What is the approximate weight of the director? *A*. 725 pounds.

Q. What is the method of prediction? *A*. Linear speed method.

Q. Is electric power necessary for the operation of the director? *A*. Yes. The constant speed motor and the prediction motors require 110-volt, 60-cycle, alternating current, for operation. In addition the data transmission system requires 110 volts alternating current.

Q. When the director is used for low-flying land or sea targets, what changes are necessary in the duties of the horizontal range setter and altitude setter? *A.* The altitude setter matches the ϵ_o dials, using the altitude handwheel. The horizontal range setter, using the range handwheel, matches the horizontal fire indices on the altitude dial after verifying that the horizontal range counter agrees with the range being read at the height finder.

Q. Can the director be operated from an offset position? *A.* Yes; the offset must not exceed 450 yards in the north-south or east-west direction. The corrections for the offset of the director must be set

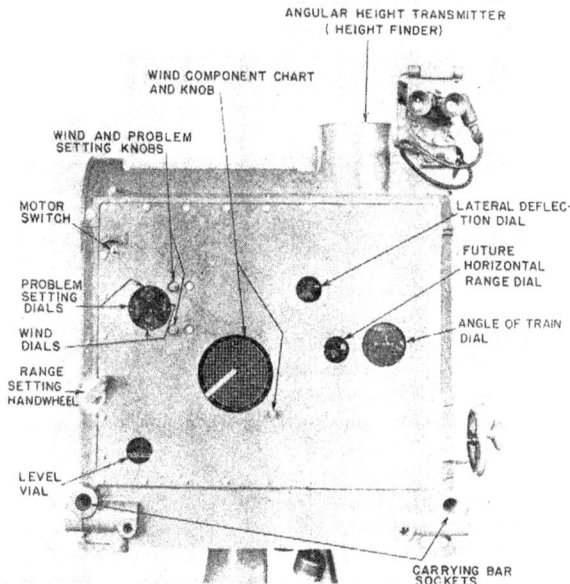

FIGURE 55.—M4 director (left side).

on the parallax dials. Vertical parallax can be set as well as horizontal.

Q. What happens when the red button is pressed in? *A.* When pressed in, the red button opens the electrical circuit to the prediction motors. Hence while the red button is pressed in, the amount of the prediction cannot change as the prediction motors are not allowed to operate.

Q. Name some of the occasions when the red button would be used. *A.*

(1) If either tracker fell behind the target, the red button would be pushed in while the tracker was catching up and for 5 seconds after normal tracking was resumed.

(2) If the horizontal range setter finds that the ϵ_o dials are not matched and that it requires relatively large movement of the range handwheel to match the pointers, the red button would be pushed in during the operation of matching and for 5 seconds after.

(3) If the altitude sent to the director by the height finder changes appreciably, the red button would be pushed in while the pointers on the altitude dial were being matched and for 5 seconds after.

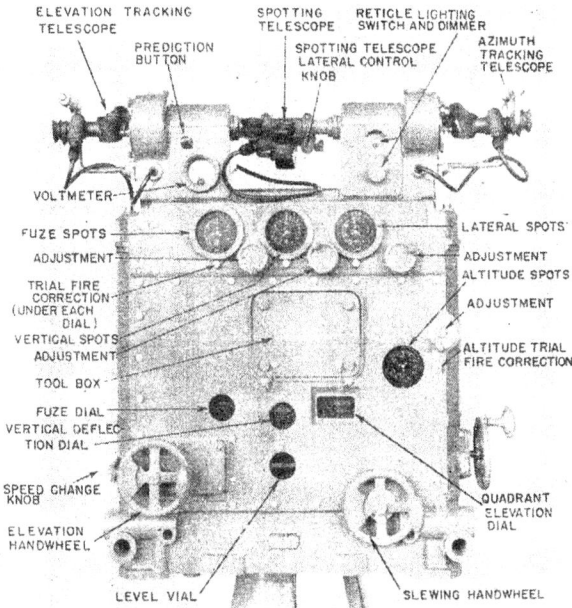

FIGURE 56.—M4 director (rear).

(4) When an altitude spot is set in the director, the red button is pressed in while the spot is being placed in the director and for 5 seconds after.

Q. What safety devices are incorporated in the director to prevent damage to the mechanism? *A.* Certain handwheels are provided with friction clutches which function when the handwheel reaches a stop. Other handwheels are equipped with positive stops. The future position slides operate a switch in the electrical circuit of the prediction motors. The motors are turned off whenever the minimum or maximum future horizontal range is reached.

Q. How is the altitude prediction mechanism used? *A.* The altitude setter turns on the altitude prediction switch (located underneath the altitude handwheel on the base plate). A rate of change of altitude is set on the altitude rate dial. As the height finder continues to send new altitudes, the altitude setter notes whether the pointers on the altitude dial remain approximately matched. If they do not, he changes the altitude rate until they do.

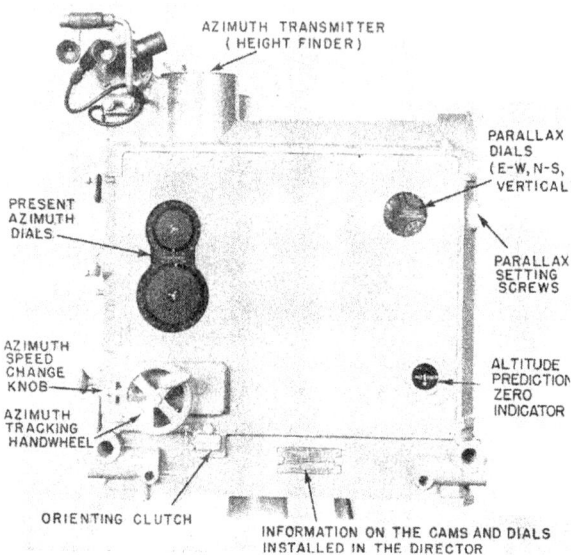

FIGURE 57.—M4 director (right side).

Q. How is the director oriented? *A.*

(1) Level the director.

(2) Remove the cover of the orienting clutch (right side of director) and turn the clutch lever to the IN position.

(3) Turn the azimuth tracking handwheel until the present azimuth dials read the azimuth to the orienting point.

(4) Turn the orienting clutch lever to the OUT position.

(5) Rotate the director by means of the sluing handwheel until the vertical cross hair of the azimuth telescope is on the orienting point.

(6) Turn the orienting clutch lever to the IN position.

(7) Check to see that the present azimuth dial still reads the azimuth of the orienting point.

(8) Replace the cover over the orienting clutch.

Q. How is a correction for wind applied in the director? *A.* Rotate the radial arm of the wind component solver (left side of director) to indicate the azimuth of the ballistic wind as given in the meteorological message. At the point on the radial arm reading the

FIGURE 58.—M3 director (front and right).

velocity of the ballistic wind as given in the meteorological message, read the N-S and E-W components of the wind. The components are then set on the wind dials. For example: A ballistic wind of 20 mph velocity and azimuth 2,400 mils would have a south component of 14 mph and an east component of 14 mph.

33. Directors M3 and M3A1.—The same questions apply to both directors. The essential difference in the directors is that the M3A1 is equipped to handle the greater range of the 105-mm gun.

Q. What method of prediction is used in this director? *A.* The linear speed method.

Q. What case firing is used with this director? *A.* Case III only. That is, the director furnishes the angle of train, quadrant elevation, and fuze range to the guns.

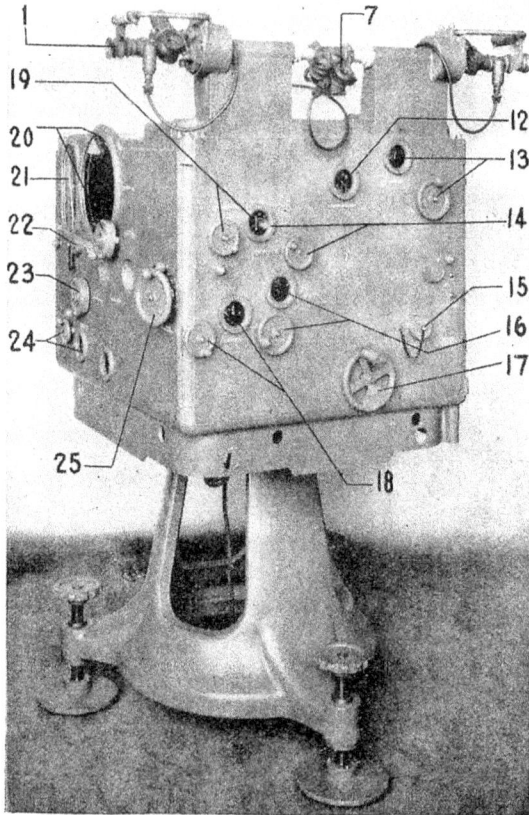

FIGURE 59.—M3 director (rear and left).

Q. Can the director be used against land or water targets? *A.* Yes. In this case the height finder furnishes horizontal ranges to the director instead of altitudes.

Q. What changes in drill are necessary when horizontal fire is used instead of antiaircraft fire? *A.* The horizontal range setter and the altitude setter continue to operate the same handwheels respectively but exchange dials to be matched. That is, the altitude setter matches the present angular height dials, using the altitude handwheel. The horizontal range setter matches the pointer on the altitude dial, using the horizontal range handwheel and range-rate handwheel.

Q. How is the director changed to operate in either horizontal fire or antiaircraft fire? A. An altitude of 3,000 to 5,000 yards is set on the altitude dial. The horizontal range handwheel is operated until the present angular height dials indicate 800 mils. The clutch for horizontal or antiaircraft fire is now turned to the desired position.

NOTE.—In order to shift the clutch from one position to another it may be necessary to manipulate the horizontal range handwheel slightly.

Q. Can the director be operated from an offset position? A. Yes.

Q. How is it set for operation from an offset position? A. Remove the parallax dial cover plate and set in the parallax corrections. These corrections are the N-S and E-W components of the displacement of the director from the battery directing point.

Q. How is the director oriented? A.

(1) Level the instrument.

(2) Set the azimuth of the orienting point on the present azimuth dials.

(3) Disengage the orienting clutch. Traverse the director until the vertical cross hair of the azimuth tracking telescope is on the orienting point. Reengage the orienting clutch.

(4) Check by traversing the director off the orienting point and then coming back again.

Q. What personnel errors will cause the director to send out poor data? A.

(1) Uneven tracking by the trackers will cause uneven rates to be computed by the rate setters. Also, the horizontal range setter will not be able to match the pointers on the present angular height dials evenly. Both of these conditions will cause poor data to be computed. If either the altitude or the horizontal range setter does not match his dials accurately and smoothly, the accuracy and smoothness of the data put out will be affected.

(2) If the rate setters constantly change their rates or operate the tachometers improperly the data output of the director cannot be smooth and accurate. In general, any inattention or improper functioning on the part of any of the director crew will result in bad data.

Q. What is the weight of the director? A. About 800 pounds.

131

Q. What power is necessary for the operation of the director? *A.* 110-volt, 60-cycle alternating current is necessary to operate the prediction motors, the range-rate motor, and the data transmission system.

Q. How is the ballistic wind applied to the director? *A.* The N-S and E-W components of the ballistic wind are determined, using the wind component solver and data from the meteorological message. The cover over the tachometers is slid to one side. The indices are displaced on the tachometers according to the amounts of the wind components. For example, a ballistic wind from azimuth 2.100 and velocity 30 mph would have an east component of 28 mph and a south component of 15 mph.

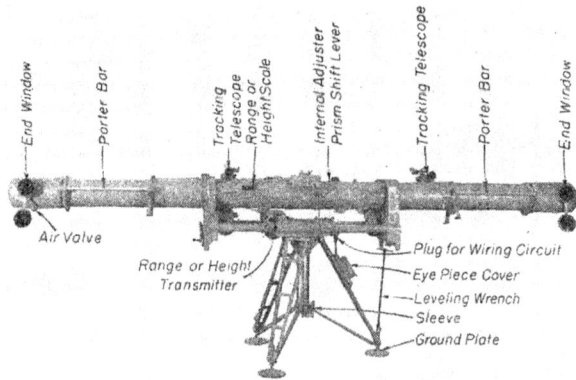

FIGURE 60.—Height finder M1 (front).

34. Height finders M1 and M2.—*a. M1.*—*Q.* What manning detail is required for the operation of the height finder? *A.* The height finder sergeant, the stereoscopic observer, the elevation tracker, the azimuth tracker, and the height or range setter.

Q. How is the height finder disassembled for transporting? *A.* It is broken down into three parts, each of which is packed separately in a steel chest or tube. The three parts are the tube, cradle, and tripod. The weight of the unit packed is 2.160 pounds.

Q. How is the tripod set up? *A.* The legs are spread out to the mechanical limit or as far as the ground will permit and the sleeve clamped in place. The tripod is approximately leveled by adjusting the ground plates. Usually two men are sufficient to set up the tripod.

Q. How is the cradle placed on the tripod and leveled? *A.* Two men, one at each end, place the cradle on the tripod so that the locking bolts engage their respective holes in the tripod top. The locking bolts are then turned down. The level of the cradle is checked by means of the level bubbles mounted on it. The accurate leveling of the cradle is done by adjusting the tripod ground plates.

Q. How is the tube mounted on the cradle? *A.* Eight men mount the tube on the cradle. The hinged end of the cylindrical carrying case is opened and the tube is pulled out after the carriage clamps on the inside wall of the case are released. The tube is then removed from the traveling carriage by loosening the lock screws at each end and shifting the clamping brackets. Before placing the tube on the cradle, the elevation receiver should be set to read zero. The elevation index should be set to match the zero line on the tube. If the index and the zero line do not match, the elevation drive can be turned

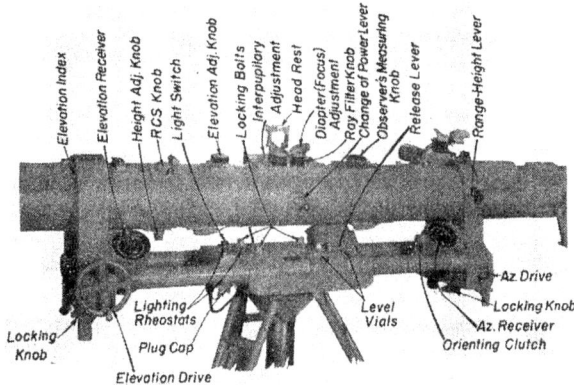

FIGURE 61.— Height finder M1 (rear center).

by the fingers where it meshes with the elevation drive of the cradle until they match. Note that the elevation receiver should be zero and the index matched with the zero line whenever the instrument is disassembled. The tube is now lifted carefully by means of the porter bars and seated in place on the cradle. When the tube is accurately seated, the locking knobs are turned down. The plug for the lighting circuit is engaged with the socket on the cradle. The light switch is turned to the proper setting depending upon the source of power. All lighting fixtures should be checked for operation.

Q. How is the height finder oriented? *A.*

(1) Level the instrument.

(2) Traverse the height finder until the center symbol of the reticle is accurately centered on the orienting point.

(3) Remove the cover on the orienting clutch on the azimuth receiver. Turn the clutch until the mechanical (outer) dial indicates the azimuth of the orienting point. Replace the cover.

Q. Why is smooth tracking so important in the height finder? *A.* The height finder proper magnifies three times as much as the tracking telescope. Hence a slight unevenness in tracking will be three times as great to the stereoscopic observer as it appears to the trackers. This results in flickering of the target over the field, thus making stereoscopic contact difficult or even impossible to obtain.

Q. What two methods are there for an observer to establish stereoscopic contact when reading ranges or altitudes? *A.*

(1) The first method, in which the observer breaks contact between successive readings, is the method most used and which should be used by all observers in the early period of training. In obtaining stereoscopic contact by this method, the observer should oscillate the range knob so that the target appears to pass in front of and then in rear of the main reticle symbol, thus obtaining a stereoscopic bracket. He then reduces the amount of oscillation until finally the target comes to rest in the same apparent plane as the main symbol, where stereoscopic contact is obtained. The observer then calls, "Read," and the reader reads the range or altitude. Then contact is broken by the observer and remade as before.

(2) In the second method, the observer after once establishing stereoscopic contact maintains it throughout. This method must be used where continuous altitudes are necessary, such as on a diving target.

Q. How are the exposed optical surfaces of the height finder cleaned? *A.* Dust is removed with the camel's-hair brush furnished with each instrument. Moisture is removed, using the selvyt cleaning cloth. Grease is removed, using the selvyt cloth moistened with alcohol. Extreme caution should be exercised to avoid scratching the glass.

b. M2.—Q. What manning detail is required for the operation of the M2 height finder? *A.* The height finder sergeant, the stereoscopic observer, the elevation tracker, the azimuth tracker, and the height or range setter.

Q. How is the height finder disassembled for transporting? *A.* It is broken down into three parts, each of which is packed in a steel chest or tube. The three parts are: tube, cradle, and tripod.

Q. How is the tripod set up? *A*. The legs are spread out to the mechanical limit or as far as the ground will permit, and the sleeve clamped in place. The tripod is approximately leveled by adjusting the ground plates. Usually, two men are sufficient to set up the tripod.

Q. How is the cradle placed on the tripod? *A*. In placing the cradle on the tripod, it should be so placed that the three index marks on the tripod match the similar marks on the cradle. The cradle is then secured to the tripod by means of the six tripod clamp screws. Although two men may be used for this operation, it is preferable that four men, one for each transportation handle, be utilized.

Q. How is the tube mounted on the cradle? *A*. The tube proper is removed from its carrying case and carriage in a manner similar to that described for the M1 height finder (*a* above). It should then be mounted on the cradle by a screw of eight men under the supervision of the height finder observer. Lifting should be done only on the porter bars provided for that purpose. The tube should first be set on the cradle so that the mounting blocks on the tube rest partially on the left half of the corresponding blocks on the cradle. Then, while the cradle and tripod are being held steady so as not to upset, the tube should be slid forward (away from the observer's side of the instrument) so that the machined shoulders of the corresponding mounting blocks are together. Keeping these surfaces together, the tube is then slid to the right until the right rear mounting block on the tube comes firmly against the locating pin on the corresponding cradle mounting block.

Q. How is the height finder oriented? *A*. The instrument is so constructed that it is oriented when it is leveled. To insure proper leveling, the instrument should be checked for level throughout 360° of traverse. To orient in azimuth, traverse the height finder until the center symbol of the reticle is accurately centered on the orienting point. Remove the cover on the orienting clutch on the azimuth receiver. Turn the clutch until the mechanical (outer) dial indicates the azimuth of the orienting point. Replace the cover.

Q. How do the azimuth and elevation tracking mechanisms on the M2 height finder differ from those on the M1 height finder? *A*. On the M1 height finder the azimuth tracking mechanism is divided between the cradle and the tripod, while the elevation tracking mechanism is divided between the cradle and the tube assembly. On the M2 height finder this division of tracking mechanism is avoided, all mechanism pertaining to azimuth tracking being contained in the cradle and that pertaining to elevation tracking being incorporated in the tube assembly proper.

35. Altimeter M1920.—*Q.* How is the altimeter set up? *A.* The tripod is set up at the position designated as the station. The height of the legs is adjusted so that the instrument will be at a convenient height for operation. The reader places the stem of the spindle bushing support in the tripod head and tightens the wing nut. He loosens the leveling screws to prevent the spindle bushing striking them. The observer places the instrument on the tripod, inserting the spindle bushing into the spindle bushing support. He holds the instrument as nearly vertical as possible while the reader tightens the leveling screws. The reader loosens the azimuth clamp and turns the instrument until one of the level bubbles is parallel to the line joining two leveling screws. He turns the two leveling screws simultaneously until the level bubble is centered. The other bubble is then centered by turning the third leveling screw. In order to do this it will be necessary to loosen or tighten the first two leveling screws by an equal amount. After the bubbles are leveled the instrument is turned through 180° and checked for level.

Q. How many men are necessary to operate the altimeters? *A.* A noncommissioned officer and three men as follows:

(1) *No. 1.*—The B^1 reader, who is the noncommissioned officer in charge of the altimeter. He is equipped with a head set and takes post seated and facing the curve disk of the B^1 instrument.

(2) *No. 2.*—The B^1 observer, who takes post at the sight of the B^1 instrument.

(3) *No. 3.*—The B^2 reader, who is equipped with a head set and takes post seated and facing the curve disk of the B^2 instrument.

(4) *No. 4.*—The B^2 observer, who takes post at the sight of the B^2 instrument.

Q. How is the altimeter oriented? *A.* By one of the three following methods:

(1) *Stations intervisible.*—(*a*) B^1 *instrument.*

    1. The B^1 observer inserts the locking pin in the telescope support so that the sight points in the same direction as the arrow on the curve disk.

    2. He releases the azimuth clamp nut and rotates the instrument until the B^2 instrument is on the vertical cross hair. He then tightens the azimuth clamp nut, and removes the locking pin.

(*b*) *B*² *instrument.*

    1. The *B*² observer inserts the locking pin in the telescope support so that the sight points in the same direction as the arrow on the curve disk.

    2. He loosens the azimuth clamp nut and sets the azimuth at 3,200. He then tightens the azimuth clamp nut.

    3. He loosens one leveling screw and turns the entire instrument about until the *B*¹ instrument is on the vertical cross hair. He then tightens the leveling screw, checks the elevel of the instrument, and verifies the sighting on the *B*¹ instrument.

    4. He then loosens the azimuth clamp nut, sets the azimuth at zero, tightens the azimuth clamp nut, and removes the locking pin.

(2) *By known azimuths. B*¹ *and B*² *instruments.*—(*a*) The observer inserts the locking pin as previously described.

(*b*) He loosens the azimuth clamp nut, sets off the azimuth of the datum point, and tightens the azimuth clamp nut.

(*c*) He then loosens one leveling screw and turns the entire instrument about until the datum point is on the vertical cross hair. He then tightens the leveling screw, checks the level of the instrument, and verifies the sighting on the datum point.

(*d*) He loosens the azimuth clamp nut, sets off the azimuth of the base line, and tightens the azimuth clamp nut. The azimuth for both instruments is the azimuth of the base line from *B*¹ to *B*².

(3) *By compass. B*¹ *and B*² *instruments.*—(*a*) The observer sets off the proper magnetic declination on the small scale near the end of the declinator.

(*b*) He loosens the azimuth clamp nut, sets the azimuth at zero, and tightens the azimuth clamp nut.

(*c*) He loosens one leveling screw and turns the entire instrument until the compass needle rests opposite its index. He then tightens the leveling screw and checks the level of the instrument.

(*d*) He loosens the azimuth clamp nut, sets off the azimuth of the base line, and tightens the azimuth clamp nut.

(This method of orientation is not accurate and should be used only when no other method is available.)

Q. How may the orientation of each instrument be verified? *A.* The orientation may be verified by taking readings at the same instant on the sun or some other celestial body. The angles read on the curve disk of each instrument should be the same if the instruments are properly oriented.

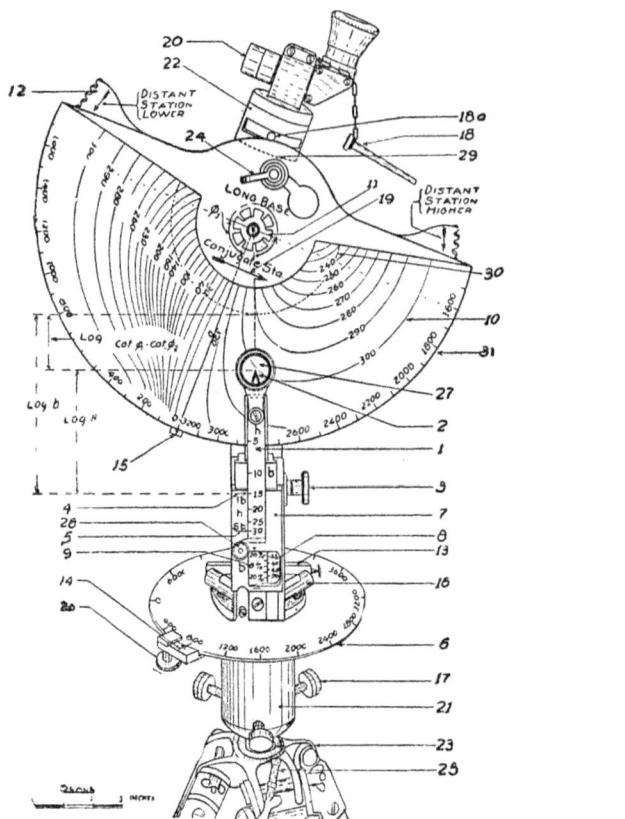

FIGURE 62.—Altimeter M1920.

1. Altitude scale (H) (numbered in hundreds of yards).
2. Altitude scale pointer (ϕ_2).
3. Altitude scale pointer knob.
4. Altitude index (lb), long base.
5. Altitude index (sb), short base.
6. Azimuth circle.
7. Base line and ballistic corrections slide.
8. Base line scale (b).
9. Base line index (0%).
10. Curve disk, ϕ_2 angles. Reverse side short base.
11. Curve disk retainer.
12. Curve disk support.
13. Declinator.
14. Azimuth index.
15. Elevation compensating index (slope).
16. Level.
17. Leveling screw.
18. Locking pin.
18a. Locking pin hole.
19. Orienting arrow.
20. Sight.
21. Spindle bushing support.
22. Telescope support.
23. Tripod head.
24. Wing nut (curve disk).
25. Spindle bushing support clamp screw.
26. Azimuth clamp nut.
27. Magnifying glass.
28. Base line slide clamp screw.
29. Line of separation between telescope supports, upper and lower.
30. Reference circle ($H = b$).
31. ϕ_1 or ϕ_2 graduated in 10 mils.

Q. Where is the B^1 instrument located? *A.* It is located at or near the battery.

Q. How is the altimeter prepared for operation? *A.*

(1) B^1 *instrument.*—(*a*) The reader sees that the "long base" or "short base" curve disk is set on the instrument as ordered by the battery commander.

(*b*) He moves the elevation compensation disk until zero is opposite the index unless otherwise directed by the battery commander.

(*c*) He sets the value of the altitude correction in percent on the correction slide opposite the length of the base line as directed by

FIGURE 63.—Gasoline-electric a-c generating unit M4.

the battery commander. If no altitude correction is ordered he sets the zero on the correction slide opposite the length of the base line.

(2) B^2 *instrument.*—The reader sets the elevation compensating index as described for the B^1 instrument. It should be remembered that the words "distant station" on the instrument refer to B^2.

Q. Explain how altitudes are obtained using this system. *A.* B^1 and B^2 observers track the target. B^2 with his altitude scale completely depressed notes the graduation on the edge of the curve disk which will be under the pointer in about 5 seconds, for example 150 and warns B^1 "Ready for 150" (one-five-zero). He again warns "Ready" when the graduation is almost under the pointer, and "Take" when the two coincide. B^1 reader elevates the altitude scale pointer until it coincides with the 150 curve on the disk. He continues following the curve until "Take" is given by the B^2 reader, at which he holds the pointer motionless and reads the altitude of the target on the altitude scale opposite the index corresponding to the face of the disk in use (sb or lb). He transmits this to the director, as for example "Altitude, twenty-two fifty."

36. Power plant and data transmission system.—Q. What is the rating of the power plant? A. The power plant is a gas-driven

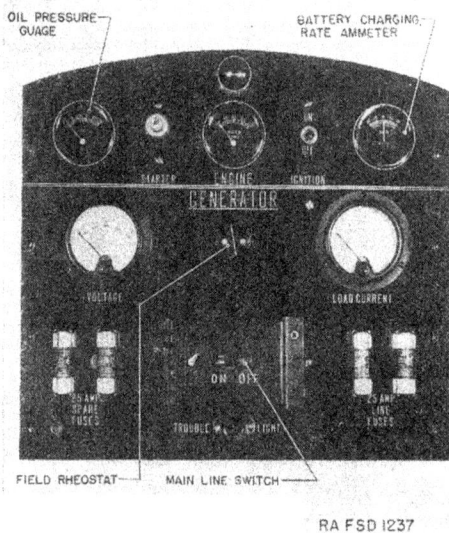

FIGURE 64.—Instrument panel.

generator. It is rated as 2½ kva, 125 volts, 60-cycle alternating current and operates at 1,200 rpm.

Q. What care must be given the power plant? A. The gas engine must have gasoline, oil, and water. It requires the same care and attention that an automobile engine would receive. The storage battery needs periodic attention.

Q. How is the power plant operated? A. With the circuit breaker in the "off" position, turn the ignition switch on. Press the starter button until the motor starts to run. Build up the voltage to approximately 110 by turning the field rheostat. Throw the circuit breaker to the "on" position.

Q. What does the agostat do? *A.* The agostat is a time delay switch. At a time from 2 to 10 seconds after the circuit breaker is thrown on, a resistance is cut out of the circuit and the full 110 volts are then impressed on the system. While the agostat is functioning the line voltage is reduced to about 70 volts. In later models the agostat is replaced by a time-delay relay operated by a small motor.

Q. Can a post power system or other source of power be used instead of the power plant? *A.* Yes, provided that it is 110-volt, 60-cycle alternating current. Special connections must be made inasmuch as terminal strips 4, 5, and 5A in the main junction box must be powered. This should only be done under the immediate supervision of an officer.

Q. What cables must be hooked up? *A.*

(1) Cable from the power plant to the main junction box.

(2) Cable from the director to the main junction box.

(3) Cables from each gun to the main junction box.

(4) Cable from the height finder to the director.

Q. What is meant by interchangeability of the cables? *A.* Any cable can be used between the main junction box and any other piece of equipment, or between the director and the height finder. All cables have 19 pole receptacles which can be plugged in anywhere in the system. *Never run a cable direct from the power plant to the director.*

Q. Operate the power plant. *A.* Practical demonstration.

Q. Hook up the cables of the system. *A.* Practical demonstration.

37. Flank spotting instrument M1 and flank spotting rule M1.—*Q.* What are the duties of the detail in operating the flank spotting instrument M1? *A.*

(1) The setting up and orienting of the instrument at the designated station.

(2) The tracking of the target.

(3) The reading and transmission of ϵ_m and θ to the battery station.

(4) The measurement of the angular deviation of the bursts from the flank and their transmission to the battery station.

Q. How is the flank spotting instrument M1 set up? *A.* The tripod is set up and leveled. The base is then screwed into position. The instrument is placed on the vertical shaft of the base. The elbow telescope is placed in the bracket and the clamp screws tightened.

Q. How is the flank spotting instrument M1 oriented? *A.*

(1) Set the ϵ_m scale so that the pointer is opposite zero. Lock in position with clamping screw.

(2) Rotate the instrument on the vertical shaft of the base until the pointer "To battery" on the θ scale is directed toward the O_1 (battery) station.

(3) Tighten the clamping knob.

(4) Loosen the ϵ_m scale clamping screw.

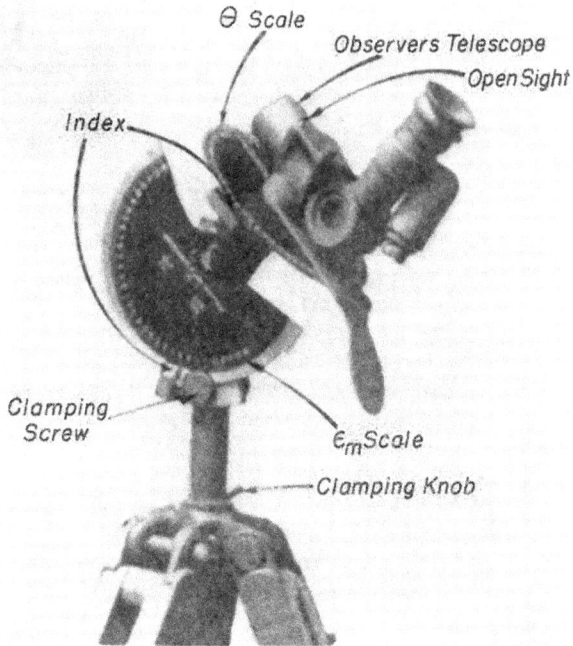

FIGURE 65.—Flank spotting instrument M1.

Q. What elements of data are obtained from the flank spotting instrument M1? *A.*

(1) The deviation of the burst in mils.

(2) ϵ_m, which is the angle made by the slant plane, containing the line of sight and the base line, with the horizontal.

(3) θ, which is the acute angle measured in the slant plane, that the line of sight makes with the base line.

Q. What elements of data must be furnished in order to operate the flank spotting rule M1? *A.*

(1) b, the length of base line O_1 to O_2 in yards.

(2) H, the altitude of the target in yards.

(3) ϵ_m.

(4) θ.

(5) The deviations of the bursts from the target in mils.

Q. How is the flank spotting rule M1 used? A. The rule consists of five scales, two of which are on movable, concentric rings. The movable scales have studs inserted in them to assist in positioning them. A movable xylonite plate with a scale engraved on it fits

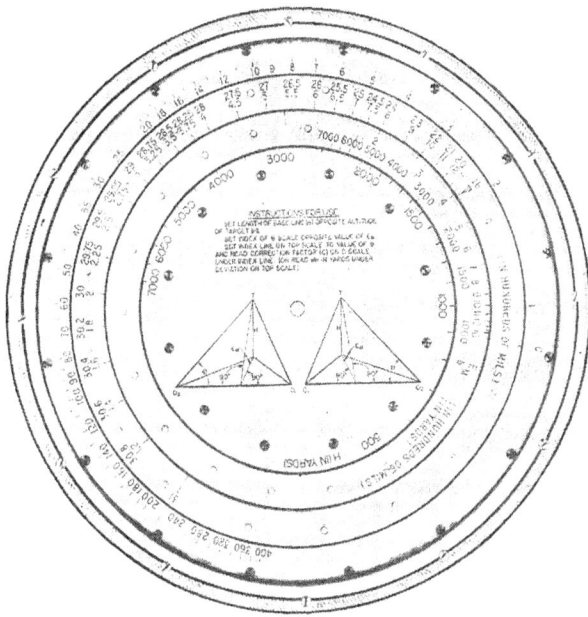

FIGURE 66.—Flank spotting rule M1.

over the entire rule. The base line b is set opposite the altitude of the target H. The index of the θ scale is set opposite the value of ϵ_m. Opposite the value of θ read the value of the conversion factor C on the outer scale. Turn the movable xylonite plate so that l is over the value of the conversion factor C. Under the deviation in mils on this scale read the altitude correction in yards on the scale marked C (outer scale).

38. Antiaircraft BC observation instrument M1.—Q. What are the principal uses of the AA BC observation instrument? A. It enables the battery commander to study and identify targets. It

143

permits him to follow closely the firing on a target. It can be used to determine the deviations of bursts when firing at a fixed point. It may be used as a transit for orientation.

Q. How is the instrument packed for transportation? *A.* It is packed in three parts. The telescope, the mount, and the tripod, after being folded compactly and strapped, are each packed separately in wooden chests.

Q. How is the instrument set up? *A.* The legs of the tripod are well spread to give stability. The legs are pushed well into the ground while the head of the tripod is kept approximately level. The mount is screwed firmly on the tripod, the cap screws of the trunnion supports unscrewed, and the trunnion supports opened to receive the trunnions of the telescope unit. The telescope unit is placed carefully in the trunnion bed. The trunnion cap squares are then folded down and the trunnion clamp screws tightened.

Q. How is the instrument oriented? *A.*

(1) Set the main azimuth scale and vernier to read the azimuth of the orienting point.

(2) Loosen the leveling screws, grasp the mount above the leveling screws and rotate the assembly until the telescope is pointed approximately at the orienting point.

(3) Tighten the leveling screws and level the instrument.

(4) Release the azimuth release clamp and center the telescope on the point.

(5) Tighten the azimuth release clamp, finally center exactly by means of the tangent screw.

Q. What is the smallest reading of the elevation scales? *A.* The main elevation scale has graduations every 10 mils. The elevation vernier has a least reading of 1 mil.

Q. How is the azimuth scale graduated? *A.* The main scale is graduated in units of 20 mils, while the least reading of the vernier is 0.1 mil.

Q. How should the traversing handwheel be located with respect to the observer's eyepiece? *A.* The traversing handwheel should be set to the right of the observer's eyepiece. To do this release the azimuth release clamp and turn the telescope to the proper position. Tighten azimuth release clamp. This must be done before orienting.

Q. What happens to the cross hairs in the telescopes as the instrument is elevated? *A.* They become inclined. Above 800 mils the horizontal cross hair is closer to the vertical. Care must be taken when observing that the vertical and horizontal cross hairs and scales are not confused.

Q. How is the telescope illuminated at night? *A.* A 4-volt battery is provided for illuminating the azimuth and elevation scales, observer's cross wires, battery commander's cross wires, and mil graduations. The amount of light is regulated by a cartridge rheostat on the base of the mount.

Q. What care should be given the instrument? *A.* The optical surfaces should be kept clean. Dust should be removed with a

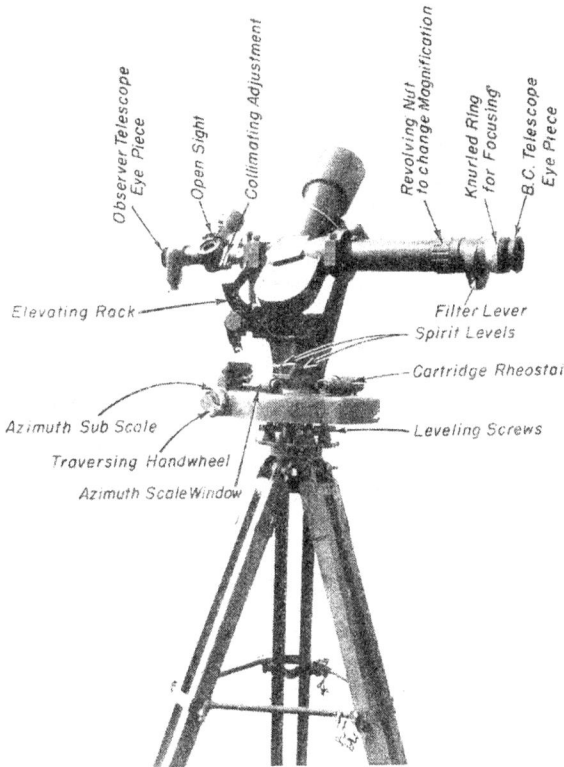

FIGURE 67.—Observation instrument, AA BC, M1 (BC telescope).

camel's-hair brush or optical paper. Moisture or grease should be removed with optical paper moistened with alcohol. Scratching the lenses should be avoided. This is a precision instrument, and it must be handled carefully. The instrument should not be unnecessarily exposed to moisture or dust. In leveling the instrument the leveling screws should not be forced if they bind.

Q. What is the procedure when the observer is directed to observe trial or calibration fire? *A.* The instrument is set up and oriented. The instrument is then set at the azimuth and angular height designated by the range officer. When the bursts occur, the observer calls out the deviations as right or left, or above or below.

SECTION III

POINTING AND POINTING TESTS

Paragraph

Pointing and pointing tests_____ 39

39. Pointing and pointing tests.—Instruction in pointing and pointing tests will be practical and will be limited to the guns and fire-control equipment with which the organization is equipped.

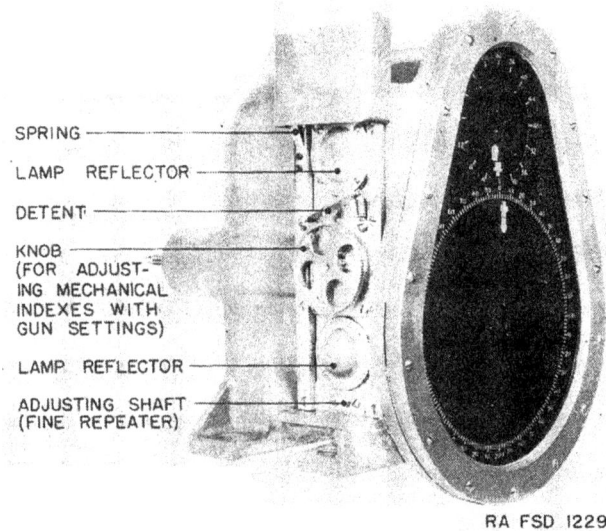

SPRING

LAMP REFLECTOR

DETENT

KNOB
(FOR ADJUST-
ING MECHANICAL
INDEXES WITH
GUN SETTINGS)

LAMP REFLECTOR

ADJUSTING SHAFT
(FINE REPEATER)

RA FSD 1229

FIGURE 68.—Azimuth (or elevation) indicator M4, left-hand cover raised.

Q. What elements of data are necessary to lay the gun, using case III firing? *A.* The angle of train, the quadrant elevation, and the fuze range.

Q. How is the gun laid according to the data? *A.* The pointers on the azimuth indicator, the elevation indicator, and the fuze range indicator are matched.

Q. Show how to match both the coarse and fine dials on the elevation (or azimuth) indicator. *A.* Practical demonstration.

Q. How is the gun oriented in azimuth? *A.*

(1) Level the gun.

(2) Using a bore sight at the breech and cross threads at the muzzle of the gun, line up the axis of the bore on the selected datum point.

(3) Raise the sliding cover on the left side of the azimuth indicator. Lift the spring detent holding the orienting knob and turn the orienting knob until the mechanical (outer) pointer indicates the azimuth

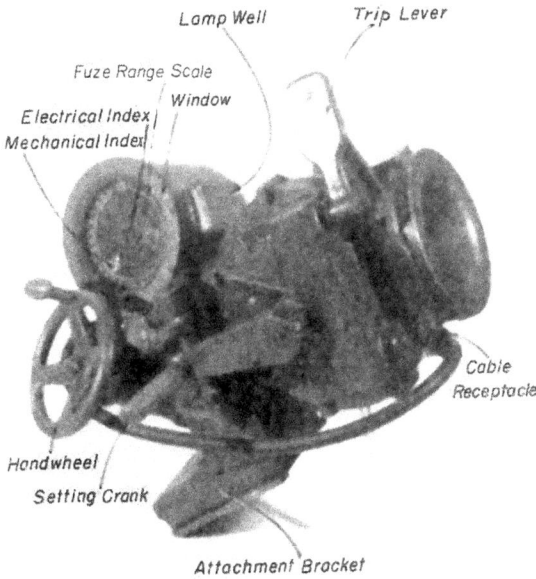

Lamp Well   Trip Lever

Fuze Range Scale
Window
Electrical Index
Mechanical Index

Cable
Receptacle

Handwheel
Setting Crank

Attachment Bracket

FIGURE 69.—M8 fuze setter.

of the datum point from the gun. Lower the detent and then the cover.

(4) Check reading and alinement of gun on datum point and readjust if necessary.

Q. How is the gun oriented in elevation? *A.*

(1) Level the gun.

(2) Set the gunner's quadrant at some convenient value, such as 700 mils.

(3) Place gunner's quadrant leveling plates on the breech of the gun and elevate the gun until the bubble in the quadrant is centered.

(4) Open the sliding cover on the left side of the elevation indicator. Raise the spring detent holding the orienting knob and turn the orienting knob until the mechanical (outer) pointer indicates the elevation as set on the gunner's quadrant.

(5) Recheck at different elevations with gunner's quadrant and reset if necessary.

Q. How is the gun oriented in fuze range? *A.*

(1) Change the setting and adjusting rings in the fuze setter to accommodate the type of fuze being used.

(2) Set the fuze indicator to some even value, such as 16 seconds, and cut a fuze.

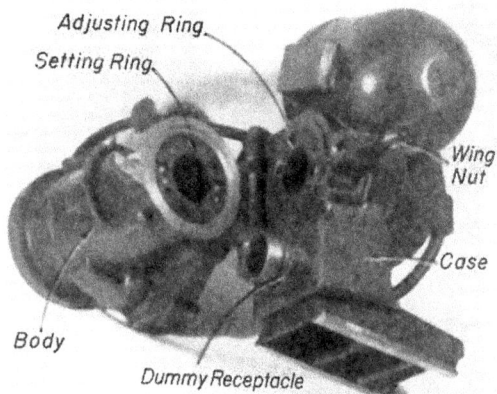

FIGURE 70.—MS fuze setter opened for inspection.

(3) If the reading on the fuze does not agree with the fuze range set on the indicator, remove the glass cover from the fuze indicator, and loosen the six clamping screws holding the clamping ring in position, thus allowing the scale to be turned.

(4) Insert a thin rod, drift, or similar object in the small hole at the top of the scale and use this to move the scale until the fuze range actually set on the projectile is opposite the mechanical (outer) pointer.

(5) Tighten the clamping screws and replace the glass cover.

(6) Set the fuze range indicator to a different value and cut another fuze to check orientation.

(7) Repeat operation if necessary.

Caution: Be sure the projectiles are all cut back to safe after using them for orientation.

Q. How are the guns synchronized? *A.*

(1) To synchronize azimuth or elevation indicators—

(*a*) Energize the data transmission system.

(*b*) Set director so that it reads any even value.

(*c*) Match pointers at guns and read value of azimuth or elevation of guns.

(*d*) If the reading on any gun differs from the value transmitted from the director, raise the sliding cover on left side of indicator on guns and turn the large screw at bottom of cavity until the electrical (inner) fine index reads the transmitted value and close the cover.

(*e*) After the indicators have been adjusted, traverse the director by a considerable amount and check to see that the indicators are still in synchronization.

(2) The fuze range indicators are synchronized with the fuze range dial on the director as follows:

(*a*) After the data transmission system is energized, set director so that fuze dial reads any even fuze range.

(*b*) Match pointers on fuze setters and check fuze readings with value transmitted by the director.

(*c*) If the readings do not agree, take off cover plate of lamp well on right side (facing it) of fuze range indicator.

(*d*) Turn the electrical synchronizing screw (found in the cavity) with a screw driver until the electrical (inner) dial reads the transmitted fuze range.

(*e*) Replace the cover plate.

(*f*) Check the indicators at several other fuze ranges.

Q. How are calibration corrections applied to the individual guns? *A.* (1) *For azimuth.*—(*a*) Set gun at some even azimuth such as 800 mils.

(*b*) Raise sliding cover on left side of azimuth indicator.

(*c*) Using toothed wheel, move mechanical (outer) fine pointer to the value of the correction in the opposite sense to that of the correction. (For example, if indicator is set at 800 mils and a plus 5 mil correction is to be applied, make pointer read 795 mils; if a minus 5 mil correction is to be applied, make pointer read 805 mils.)

(*d*) Close cover.

(2) *For elevation.*—(*a*) Set gun at any convenient elevation by means of a gunner's quadrant.

(*b*) Raise sliding cover on left side of elevation indicator and by means of the toothed wheel make (outer) mechanical dial read value of elevation plus or minus the correction value (sign reversed).

(*c*) Close cover.

NOTE.—It should be remembered that after a calibration correction has been applied to a gun, that gun will be in error by the amount of the correction in subsequent orientation in azimuth and elevation. *For this reason, the amount of the correction applied is always recorded.*

(3) *For fuze range.*—(*a*) Energize the data transmission system.

(*b*) Set director so that fuze dial reads any fuze range such as 12.

(*c*) By means of the electrical synchronizing screw (which is in lamp well on right side of fuze indicator as you face the dial), make electrical (inner) dial read fuze range plus or minus the correction. (For example, if it is desired to apply a correction of plus 2 corrector divisions, make the dial read 12.2.)

NOTE.—Remember that when synchronization is checked, the fuze setter should be out of synchronization by the amount of correction applied.

CHAPTER 6

GUNNERY AND FIRE CONTROL—AUTOMATIC WEAPONS

SECTION I

USE, ORIENTATION, AND ADJUSTMENT OF FIRE-CONTROL INSTRUMENTS

40. Use.—*Q.* What is the standard fire-control system for automatic weapons? *A.* Antiaircraft automatic gun control equipment set M1.

Q. Of what parts does the set consist? *A.* A control box, a number of 50-foot flexible shafts (cables), and the necessary packing chests for the control box and flexible shafts.

Q. Describe and demonstrate the proper method of setting up the control box. *A.*

(1) Carry the control box in its packing chest to the desired location.

(2) Release the four trunk bolts and remove the lid of the packing chest.

(3) Grasp the control box by the *base* and lift it from the chest. *Do not use the handwheels as handles in lifting or carrying the box.*

(4) Remove the split pins holding the folding seat supports to the bottom of the chest.

(5) Unfold the seats upward and outward and, with the seats approximately level and the holes in the projecting lugs of the seat supporting bars alined with the holes in the blocks at the ends of the packing chest, reinsert the split pins.

(6) Adjust the seats by sliding them in or out the desired amount.

(7) Place the control box on the blocks in front of the seats and screw the wing screws upward securely into the threaded holes in the base of the box.

Q. Describe and demonstrate the proper method of laying and connecting the flexible shafts. *A.* Remove the required number of flexible shafts from their carrying cases, being careful to avoid kink-

151

ing the shafts. Lay the cables between the control box and the guns or data computing equipment (if used). Where the distance between the control box and the other equipment is greater than 50 feet, couple the required number of flexible shafts together, using the minimum number possible. Sufficient slack must be provided to permit traversing of the guns. When directed to do so, connect the shafts to the guns and control box (data computing equipment, if used). To connect a shaft to a coupling, remove the covers from the coupling and the proper end of the shaft, aline the mating parts and force the shaft end solidly against the coupling; screw the locking ring down firmly. Whenever flexible shafts are coupled together or to a receptacle, screw the covers of the mating parts together to protect the threads from burring and from the accumulation of sand or dirt. Foreign matter allowed to collect in the cover will later work down inside the casing of the shaft and will cause binding or other difficulty.

Q. What care must be taken of the control equipment set M1? *A.*

(1) *Control box.*—(*a*) The control box should never be lifted by the handwheels.

(*b*) Ball bearings and gears should be lubricated only by competent ordnance personnel. They are lubricated at assembly and further lubrication will be required only at long intervals.

(*c*) Care should be taken that the proper lead dials and stops are being employed for the type of gun being used.

(2) *Flexible shafts.*—(*a*) Flexible shafts should not be kinked or bent sharply.

(*b*) Chafing, rubbing, or crushing of the flexible shafts should be prevented. Vehicles should not be allowed to run over them. It may be desirable to partially bury them in shallow trenches to prevent personnel from walking on them.

(*c*) The flexible shaft casing should be kept free from oil.

When the flexible shafting is connected to 37-mm guns the lead handwheels at the control box should never be turned so as to set a positive vertical lead on the gun sight unless the gun is elevated about 400 mils. (With the gun at zero elevation, an attempt to set in a positive vertical lead lowers the drive housing until it is in contact with the case or handwheel brackets. *Any further lowering of the housing will result in breaking the flexible shafting.* This warning does not apply to the M2 machine-gun mount.)

(3) *Control box and flexible shafts.*—(*a*) The ends of couplings should be kept clean and lightly oiled with neutral oil (U. S. Army Spec. No. 2–81) and the covers should be kept in place when not in use.

(*b*) Access by unauthorized personnel to the internal parts of the control box or flexible shafts for lubrication or other purposes should not be permitted.

(*c*) All parts of the control equipment set M1 should be kept in packing chests when not in use.

Q. How are leads applied at the central control box? *A*. The two adjusters, opposite the lateral and vertical adjusting knobs, set in the required leads, in mils, on the corresponding lead adjusting scales. This operation causes the corresponding lead index (lateral or vertical) to move from its normal position the same number of mils.

VERTICAL DEFLECTION

FLEXIBLE SHAFTS FROM COMPUTING DEVICE

LATERAL DEFLECTION

RA FSD 859

FIGURE 71.—Control box (input side).

Q. How are the leads applied to the guns? *A*. The lateral (or vertical) matcher, seated at the proper end of the control box, turns the corresponding lead handwheel until related transmitted lead index matches the inner index. This operation causes the flexible shaft to be rotated the correct amount to apply the desired amount of lead to the gun sights.

41. Adjustment.—*Q*. How is the control box synchronized with the guns of the fire unit? *A*. The dials of the control box are set at normal. As soon as all guns are boresighted, the required number of flexible shafts are connected from the output couplings of the control box to the proper couplings of the gun sighting systems. The system is checked by setting lead values on the lead dials and seeing that the corresponding settings are made on the gun sights. This check does three things. It checks that no dials or sighting systems were moved while connecting up the system, it discloses any defects in individual flexible shafts, and it insures that the lateral and vertical couplings of each sighting system are connected to the corresponding side (lateral or vertical) of the control box.

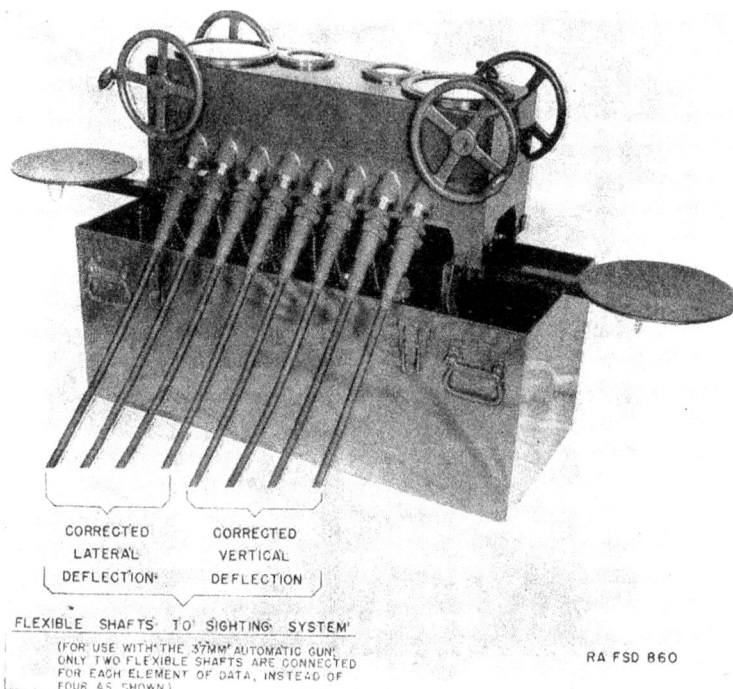

CORRECTED
LATERAL
DEFLECTION

CORRECTED
VERTICAL
DEFLECTION

FLEXIBLE SHAFTS TO SIGHTING SYSTEM

(FOR USE WITH THE 37MM AUTOMATIC GUN, ONLY TWO FLEXIBLE SHAFTS ARE CONNECTED FOR EACH ELEMENT OF DATA, INSTEAD OF FOUR AS SHOWN)

RA FSD 860

FIGURE 72.—Control box (output side).

Q. Describe and demonstrate the proper method of setting the dials of the control box at normal. A. Turn each adjusting knob until the corresponding index reads zero. Turn each lead handwheel until the corresponding transmitted lead index reads 500 (for 37-mm gun scales) or 300 (for machine-gun scales). If either lead index does not also read 500 (37-mm gun) or 300 (machine guns), remove the cover on the corresponding input coupling and turn the coupling until the proper reading is obtained. Replace the cover.

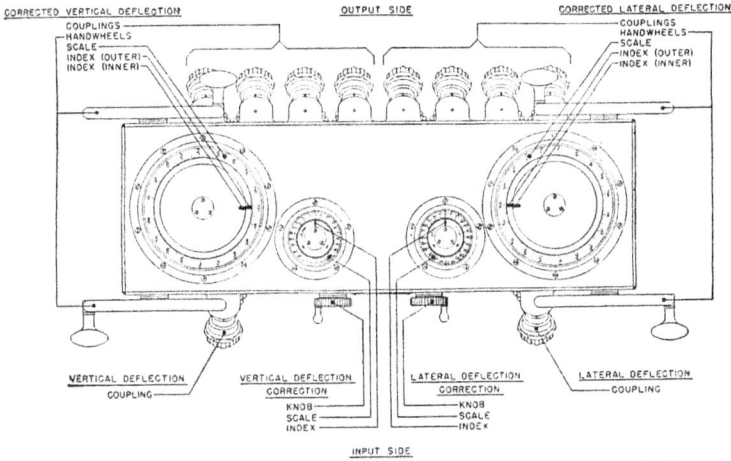

CORRECTED VERTICAL DEFLECTION
OUTPUT SIDE
CORRECTED LATERAL DEFLECTION

COUPLINGS
HANDWHEELS
SCALE
INDEX (OUTER)
INDEX (INNER)

COUPLINGS
HANDWHEELS
SCALE
INDEX (OUTER)
INDEX (INNER)

VERTICAL DEFLECTION
COUPLING

VERTICAL DEFLECTION
CORRECTION
KNOB
SCALE
INDEX

LATERAL DEFLECTION
CORRECTION
KNOB
SCALE
INDEX

LATERAL DEFLECTION
COUPLING

INPUT SIDE

RA FSO 662

FIGURE 73.—Control box (top view).

42. Drill of range section.—*Q*. What are the duties of the range section? A. See drill in table XIII.

155

TABLE XIII.—*Drill for antiaircraft automatic gun-control equipment set M1*

Details	PREPARE FOR ACTION	DETAILS, POSTS	TARGET	CEASE TRACKING	MARCH ORDER
Instrument corporal.	Repeats command and supervises work of entire detail. Indicates position of control box. Assisted by squad leader or gun commander of each gun, synchronizes control box with guns before flexible shafts are attached to guns. After shafts are attached, verifies synchronization. When range section is ready to function, reports to platoon commander, "Sir, range section in order."	Has no fixed post. Repeats the command and moves around box so that he may supervise work of entire section.	Repeats the command, giving any additional instructions necessary to aid the adjusters and spotters in quickly picking up the target.	Repeats the command.	Repeats the command and supervises work of placing all equipment in proper chests. Replaces equipment as directed.
No. 1 (vertical adjuster).	Assisted by No. 2, removes control box from packing case and sets up control box at position indicated by instrument corporal. Connects flexible shafts to vertical output couplings of control box. Tests functioning of vertical adjusting knob. Sets vertical lead adjusting index to zero. When No. 3 has set the vertical transmitted lead index at normal (500 for 37-mm gun and 300 for machine guns) and while the vertical lead adjusting index is still at zero, checks to see that the vertical lead index matches the vertical transmitted lead index. If it does not, removes cover of vertical input coupling and rotates coupling until indexes are matched. Replaces coupling cover. Connects telephone to field wire to spotters.	Takes post facing vertical adjusting knob.	Repeats command to vertical spotter, identifies target. Estimates initial vertical lead and sets it on vertical lead adjusting scale. Continues to change the scale settings as required. Scale settings are changed to conform to his estimate of rate of change of leads. Observes tracers and receives vertical spotter's reports to constantly adjust estimate of proper leads.	Repeats command to vertical spotter. Ceases operation of vertical adjusting knob but remains at post. Sets vertical lead adjusting index to zero.	Disconnects vertical flexible shafts from control box. Replaces covers on vertical output couplings. Assisted by No. 2, replaces control box in packing chest. Disconnects telephone and replaces it as directed.

No. 2 (lateral adjuster).	Assists No. 1 in removing control box from packing case and setting up control box at position indicated by instrument corporal. Connects flexible shafts to lateral output couplings of control box. Tests functioning of lateral adjusting knob. Sets lateral lead adjusting index to zero. When No. 4 has set the lateral transmitted lead index at normal (500 for 37-mm gun and 300 for machine guns) and while the lateral lead adjusting index is still at zero, checks to see that the lateral lead index matches the lateral transmitted lead index. If it does not, removes cover of lateral input coupling and rotates coupling until indexes are matched. Replaces coupling cover. Connects telephone to field wire to spotters.	Takes post facing lateral adjusting knob.	Repeats command to lateral spotter. Identifies target. Estimates initial lateral lead and sets it on lateral lead adjusting scale. Continues to change the scale settings as required. Scale settings are changed to conform to his estimate of rate of change of leads. Observes tracers and receives lateral spotters' reports to constantly adjust estimate of proper leads.	Repeats command to lateral spotter. Ceases operation of lateral adjusting knob but remains at post. Sets lateral lead adjusting index to zero.	Disconnects lateral flexible shafts from control box. Replaces covers on lateral output couplings. Assists No. 1 in replacing control box in packing chest. Disconnects telephone and replaces it as directed.
No. 3 (vertical matcher).	Removes flexible shafts from packing cases and stretches shafts for vertical leads from control box to individual guns. Connects two lengths of shafting together where necessary. Tests functioning of vertical lead handwheel. Sets vertical transmitted lead index at normal (500 for 37-mm gun and 300 for machine guns).	Takes post on seat facing vertical lead dial.	Matches indexes of vertical lead dial and carefully keeps them matched thereafter.	Ceases matching indexes of vertical lead dial, but remains at post. Sets vertical transmitted lead index at normal.	After vertical flexible shafts are disconnected from control box, replaces covers on shaft ends and then replaces shafts in packing chests.
No. 4 (lateral matcher)	Removes flexible shafts from packing cases and stretches shafts for lateral leads from control box to individual guns. Connects two lengths of shafting together where necessary. Tests functioning of lateral lead handwheel. Sets outer lateral transmitted lead index at normal (500 for 37-mm gun and 300 for machine guns).	Takes post on seat facing lateral lead dial.	Matches indexes of lateral lead dial and carefully keeps them matched thereafter.	Ceases matching indexes of lateral lead dial, but remains at post. Sets lateral transmitted lead index at normal.	After lateral flexible shafts are disconnected from control box, replaces covers on shaft ends and then replaces shafts in packing chests.

TABLE XIII.— *Drill for antiaircraft automatic gun-control equipment set M1*—Continued

Details	PREPARE FOR ACTION	DETAILS, POSTS	TARGET	CEASE TRACKING	MARCH ORDER
Nos. 5 to 10, inclusive (observers and spotters).	Receive instructions as to where their stations will be located. Secure telephones and test them. Proceed to stations, laying field wire. Upon arrival at stations, each spotter connects his telephone to the field wire and tests communication.	Remain at their telephones in the designated positions.	When fire is commenced, the designated lateral and vertical spotters observe the fire and telephone their observations to the corresponding adjuster. Other spotters prepare to observe if called upon.	Cease telephoning observations if not already discontinued. Remain at posts.	Disconnect telephones. Return to control station winding up field wire. Replace telephones as directed.
Nos. 11 and 12 (telephone operators).	Connect telephones to line to spotters (platoon commander's telepone) and to line from battery headquarters. Test communication.	Take post at telephones.	Remain at post. Relay any messages given to them by the platoon commander.	Remain at post. Relay any messages given to them by the platoon commander.	Disconnect telephones and replace them as directed. Assist in taking up field wire at platoon position as directed.

SECTION II

DETERMINATION OF FIRING DATA

43. General.—*Q.* What firing data must be determined for automatic weapons? *A.* Vertical and lateral leads. The leads applied to the guns are the angular amounts that the line of sight of each gun sight is shifted from a line parallel to the axis of the bore of its gun.

Q. What factors influence the value of the leads? *A.* Ground speed of target (S_o), minimum horizontal range (R_m), altitude of target (H), muzzle velocity of the gun, angle of dive (for diving target only), and the position of the target (T_o), along the course.

Q. In what units are the leads measured? *A.* For individual tracer control, in target lengths as that is the only unit of measure instantly available to the gunner. For central tracer control, in mils as scales, dials, and charts are prepared with this unit of measure.

Q. How are the initial firing data (initial leads) determined? *A.* They are estimated before fire is opened. After fire has been opened, adjustment of the initial leads is made based on observation of the tracers.

Q. Explain just how an adjuster determines and applies either the vertical or lateral lead. *A.* From a study of lead charts (figs. 74 and 75 and of lead characteristics, from the use of tracer trainer where available, and from actual firing the adjuster gains the knowledge of initial leads for representative type courses and of the direction and rate of change of the leads during those courses. When a target approaches, he estimates the initial lead that will be required at the point where he expects fire to be opened, based on the course and speed of the airplane, and sets this into the control box. If fire is delayed, he sets in new initial leads continuously by changing the leads in the proper direction and at the required rate. When fire is opened, he adjusts his rate of change of leads based on his own and the flank spotter's observation of the tracers. For example, if the tracer stream is moving farther and farther ahead of the target on the approaching leg (fig. 74) of a constant altitude crossing course, the lead is being increased too rapidly. The necessary adjustment of fire would be accomplished by reducing the rate at which the leads are being changed.

Q. When the target is coming directly toward the gun. what leads are needed? *A.* Only the vertical lead. The lateral lead is zero.

Q. In individual tracer control who determines the initial lead? *A.* Usually the platoon commander.

44. Lead charts.—*Q.* What is a lead chart? *A.* A representation of the required leads for one or more courses arranged conveniently for ready reference (figs. 74 and 75).

Q. To what does the lead chart in figure 74 apply? *A.* It is the lateral lead chart for caliber .50 machine guns firing M1 ammunition at targets flying on crossing courses at a constant altitude of 400 yards.

Q. What target speeds are represented? *A.* Target speeds of 50 yards per second and 70 yards per second (see heavy numerals along left edge of chart).

Q. What is meant by the minimum horizontal range (R_m)? *A.* The horizontal range to the target at the midpoint of the course (figs. 74 and 75). This is the point where the angle of approach equals 90°.

Q. What minimum horizontal ranges are represented by the courses shown in figure 74? *A.* 400, 600. 800. 1,000. and 1,200 yards. (See heavy numerals on right edge of chart.)

Q. To what does the lead chart in figure 75 apply? *A.* It is the vertical lead chart for caliber .50 machine guns firing M1 ammunition at targets flying on crossing courses at a constant altitude of 400 yards.

Q. What are the initial leads for the following courses shown in figures 74 and 75 (angle of approach 800 mils on the approaching leg)?

(1) R_m 1,000; H 400; S_g 50?
(2) R_m 600; H 400; S_g 70?
A. (1) Lateral lead—56 mils.
        Vertical lead—28 mils.
(2) Lateral lead—69 mils (or 68 mils).
        Vertical lead—35 mils.

Q. If fire is delayed until the target reaches the midpoint, what will the initial leads be for the courses described in the preceding question? *A.*
(1) Lateral lead—73 mils.
        Vertical lead—8 mils.
(2) Lateral lead—92 mils.
        Vertical lead—1 mil.

Q. From a study of the lead charts (figs. 74 and 75)—

(1) Approximately where does the maximum lateral lead occur on a constant altitude crossing course?

(2) Approximately where does the maximum positive vertical lead occur on a constant altitude crossing course?

A. (1) At the midpoint.

(2) At the beginning of the course.

Q. Considering figure 74, is the the lateral lead increasing or decreasing on—

(1) The approaching leg?

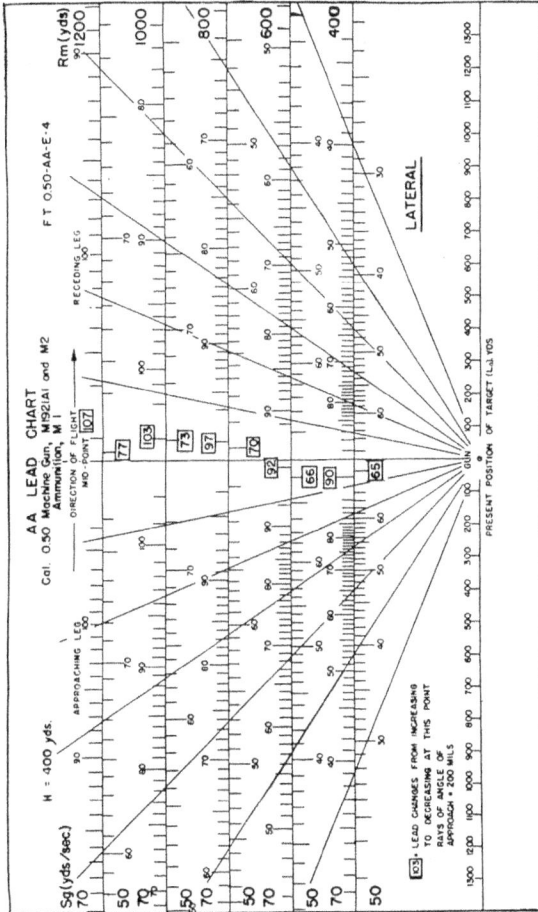

FIGURE 74.—Lateral lead chart.

(2) The receding leg?

A. (1) Increasing.

(2) Decreasing.

Q. Considering figure 75, is the vertical lead increasing or decreasing—

(1) On the approaching leg?

(2) At 400 mils past the midpoint on the receding leg?

(3) At 1,000 mils past the midpoint on the receding leg?

A. (1) Decreasing.

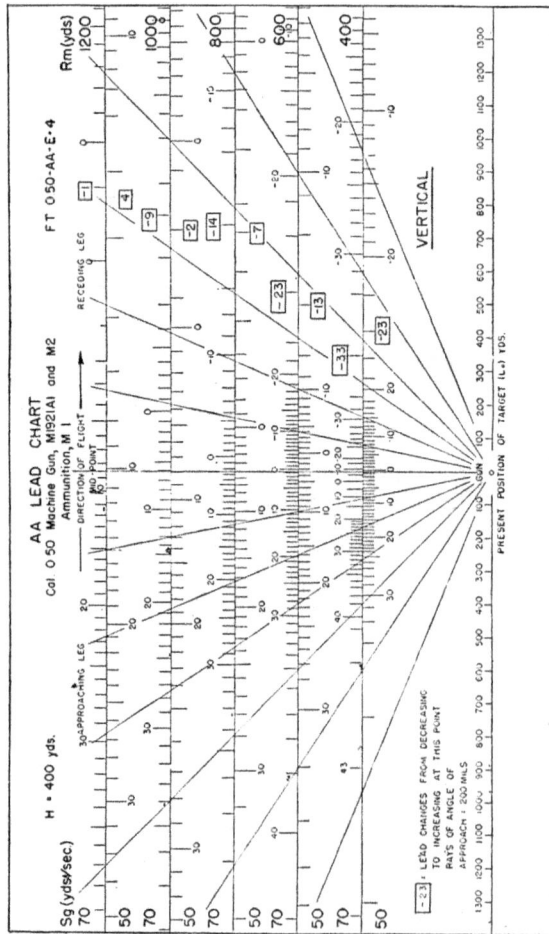

FIGURE 75.—Vertical lead chart.

(2) Decreasing.

(3) Increasing.

Q. How is the rate of change represented on the charts? *A.* By spacing of the mil divisions, which denote the lead values.

Q. For a constant altitude crossing course with R_m 1,000, H 400, and S_g 50, where does the lateral lead change least rapidly? *A.* At about the midpoint.

Q. For a constant altitude crossing course with R_m 400, H 400, S_g 70, where does the vertical lead change most rapidly? *A.* At about the midpoint. Note that graduations near the midpoint are 5-mil graduations.

Section III

CONDUCT AND CONTROL OF FIRE

45. General.—*Q.* What is the normal fire unit for automatic weapons? *A.* The platoon, consisting of either two 37-mm guns or four caliber .50 machine guns.

Q. What is the normal target for automatic weapons? *A.* Low-flying aircraft.

Q. What other targets may be engaged? *A.* Any suitable air, sea, or land targets within range. However, consideration must be given to the mission to be accomplished, the immediate threat, and the problem of ammunition supply when determining what targets will be engaged.

Q. What characteristics of automatic weapons make them more suitable than larger guns for firing at close-in targets? *A.*

(1) They are flexible, that is, they can traverse and elevate rapidly enough to keep up with a close-in target, and can be readily adapted to tracer control.

(2) They can open fire more quickly with applied data.

(3) They have a higher rate of fire.

Q. How is the fire of automatic weapons observed? *A.* By use of tracer ammunition.

Q. What are the three methods of fire control normally employed with automatic weapons? *A.*

(1) Individual tracer control (case I).

(2) Central tracer control (case I).

(3) Director with tracer adjustment (case III).

163

Q. Describe the three standard methods of conducting the fire of automatic weapons. *A.*

(1) *Individual tracer control.*—In this method each gunner adjusts his own fire by observation of the tracers from his own gun.

(2) *Central tracer control.*—In this method each gun is equipped with a sight which the gunner keeps alined with the target. The sights of all guns are controlled simultaneously from a centrally located control station by means of flexible cables. Two adjusters at the control box observe the tracers of all the guns and make the necessary changes in the positions of the sights to adjust fire. The gunners do not observe the tracers or attempt to adjust fire.

(3) *Director with tracer adjustment.*—In this method the guns are continuously and automatically trained on data transmitted from an automatic weapons director. No gun pointers are employed. Adjustment of fire is accomplished by personnel at the director, their adjustments being based on observation of tracers.

Q. What is the purpose of adjusting fire? *A.* To place the center of the cone of fire of the single gun (individual tracer control) or of the platoon (central tracer control) on the target.

Q. How is fire adjustment accomplished? *A.* By observing the deviations of the cone of fire from the target and making the necessary changes in the leads. Deviations are observed and leads changed continuously throughout the course of the target.

Q. What are the most important requirements to be met by a method of fire adjustment for automatic weapons? *A.*

(1) It must be simple in operation.

(2) It must be accurate.

(3) It must be rapid in operation.

Q. What is the general tendency as to lateral leads in firing by tracer control? *A.* The tendency is to fire behind the target, that is, to take too small a lateral lead.

Q. Explain why the tracer trajectory appears to curve to the left when firing on a left to right course but appears to curve to the right when firing on a right to left course. *A.* This apparent curvature of the trajectory is purely an optical illusion caused by observing, simultaneously, tracers fired at successive azimuths and by the movement of the airplane with respect to each tracer as it travels along the trajectory (fig. 76). The faster the airplane is traveling, the more pronounced the curve will be.

Q. How can this optical illusion be partially eliminated? *A.* The observer should center his attention on the part of the trajectory in the vicinity of the target.

46. Tracer ammunition.—*Q.* What is the advantage of using tracer ammunition for automatic weapons? *A.* Ball cartridges are not visible during flight. Tracer ammunition is visible during flight because each round has a burning compound in the base of the bullet which burns while moving along the trajectory. These bullets show the shape and path of the cone of fire up to the burn-out point so long as fire is delivered.

Q. What proportion of tracer ammunition to ball ammunition is generally used for—

(1) Machine guns?

(2) 37-mm guns?

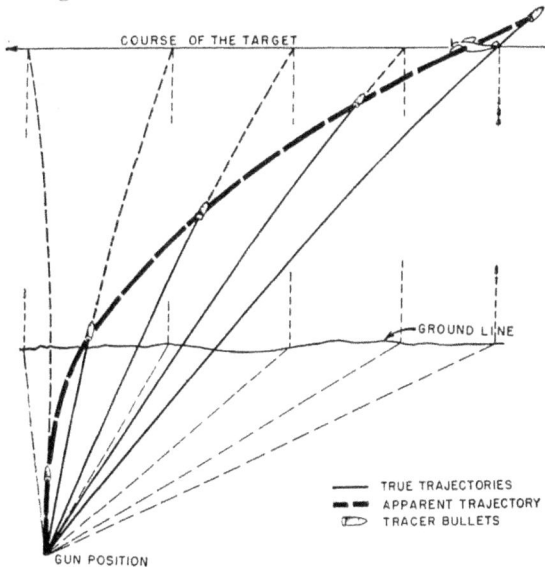

FIGURE 76.—Apparent tracer trajectory.

A. (1) One tracer to four ball cartridges, every fifth cartridge being a tracer.

(2) 37-mm ammunition is tracer ammunition.

Q. What is the approximate burn-out point of—

(1) Caliber .50 tracer ammunition?

(2) Caliber .30 tracer ammunition?

(3) 37-mm antiaircraft shell?

A. (1) 1,850 yards.

(2) 1,000 yards.

(3) 3,500 yards.

47. Individual tracer control.—*Q*. At what ranges is individual tracer control considered a satisfactory method of fire control? *A*. At ranges of less than 600 yards.

Q. In individual tracer control, who observes the firing? *A*. The individual gunner.

Q. How does the gunner point his gun in individual tracer control? *A*. He endeavors to point his gun so that the tracer stream will pass through the target selecting a point on the trajectory which he considers to be at the same range and alining this point with the target. He does not aim by sighting through the sights or over the gun except in obtaining the initial lead.

Q. With what automatic weapons may individual tracer control be used successfully? *A*. Caliber .50 and caliber .30 machine guns.

Q. Who does the adjusting in individual tracer control? *A*. The individual gunner.

48. Central tracer control.—*Q*. At what ranges is central tracer control considered a satisfactory method of fire control? *A*. At ranges less than 1,200 yards.

Q. Who observes the firing in central tracer control? *A*. The adjusters and the spotters.

Q. How many adjusters are there and where are they located? *A*. There are two adjusters, a lateral adjuster and a vertical adjuster. Both are located at the central control box.

Q. How many spotters are required for observation on one course? *A*. Two. A lateral spotter and a vertical spotter.

Q. What is the best location for the lateral spotter? *A*. From 100 to 200 yards from the firing unit along the line of fire. The shorter distances are used when observing machine-gun fire while the longer distances are used when observing 37-mm gun fire.

Q. What is the best location for the vertical spotter? *A*. From 100 to 200 yards from the firing unit on the flank from which the target is approaching. The shorter distances are used when observing machine-gun fire while the longer distances are used when observing 37-mm gun fire.

Q. What is the most satisfactory arrangement of spotters for all-around fire? *A*. Four spotters equally spaced around the gun position and 100 to 200 yards from it. When a target approaches, the spotters best situated are designated to observe fire.

Q. How are the observations of the spotters employed in adjusting fire? *A*. The spotters are connected with the adjusters by telephone. The adjusters base their adjustment of fire on their observation of fire and the reports of the spotters.

CHAPTER 7

OBSERVATION, ORIENTATION, AND RECONNAISSANCE—
HEADQUARTERS BATTERIES

SECTION I

USE, ORIENTATION, AND ADJUSTMENT OF OBSERVA-
TION INSTRUMENTS

49. Use.—*Q.* Name the different types of observation instruments employed by antiaircraft artillery units. *A.* AA BC observation instrument M1, flank spotting instrument M1, transit, and recording theodolite.

Q. What is the purpose of the AA BC observation instrument M1? *A.* This instrument is employed by battery commanders of antiaircraft gun batteries to study and identify possible targets and to follow the course of firing on a target. It is used also to determine the lateral and vertical deviations of trial shot bursts. The instrument may be employed in lieu of a transit for orienting the battery.

Q. What is the purpose of the flank spotting instrument M1? *A.* This instrument was designed to provide a simple observation instrument which would enable the flank observer of a gun battery to spot in the inclined plane containing the base line and the target. It provides a means of determining the angles ϵ_m and θ, which are used with the flank spotting rule M1.

Q. What is the purpose of the transit? *A.* The transit is employed to determine the orientation data for a gun battery. Where time is limited, map distances may be used in lieu of data determined by the use of a transit.

Q. What is the purpose of the recording theodolite? *A.* When accurate data are desired concerning the results of target practice or other firings, recording theodolites (specially constructed cameras) are employed at the battery and at the flank spotting station to record the deviations of the bursts laterally, vertically, and in range from the target. These deviations are recorded on motion-picture film along with data concerning the azimuth of the target, elevation of the target, instant of time of exposure, and serial number of the theodolite.

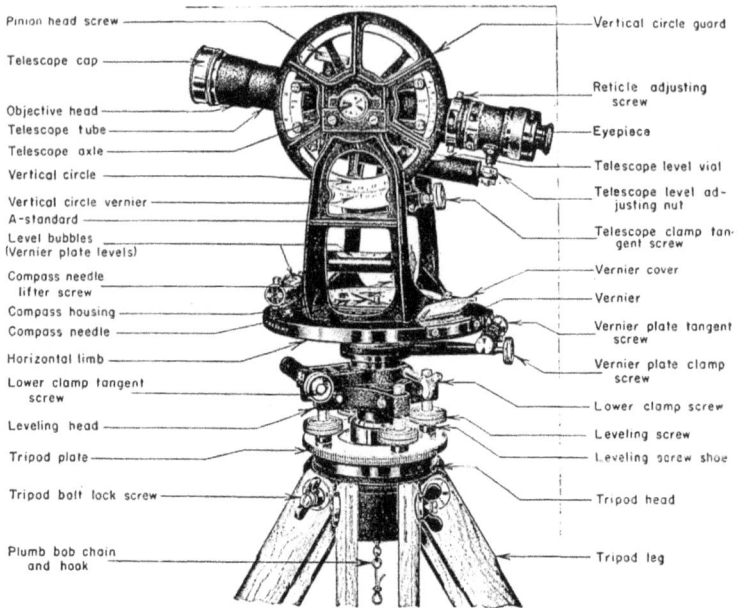

FIGURE 77.—Transit.

Q. What are the steps in setting up the **AA BC** observation instrument M1? *A.*

(1) The legs of the tripod are well spread to give stability. The legs are pushed well into the ground while the head of the tripod is kept approximately level.

FIGURE 78.—Recording theodolite.

(2) The mount is screwed firmly on the tripod.

(3) The cap screws of the trunnion supports are unscrewed and the trunnion supports opened to receive the trunnions of the telescope unit.

(4) The telescope unit is placed carefully in the trunnion bed.

(5) The trunnion cap squares are then folded down and the trunnion clamp screws tightened.

Q. By what means and when is the AA BC observation instrument M1 leveled? *A.* The instrument is leveled by means of the leveling screws and spirit levels attached to the mount. As the leveling screws are loosened when shifting the line of sight onto the orienting point leveling is not performed until the telescope unit is sighted approximately on the orienting point with the azimuth scale reading the azimuth of that point.

Q. How is the flank spotting instrument M1 set up and leveled? *A.*

(1) Set up and level the tripod.

(2) Screw the base into position.

(3) Place the instrument on the vertical shaft of the base.

(4) Insert the elbow telescope in the bracket and tighten the clamp screws.

Q. How is the transit set up and leveled? *A.*

(1) The tripod is opened and set over the selected point, with the head approximately level.

(2) The transit is fastened to the head and the plumb bob attached.

(3) The instrument is shifted until the plumb bob is directly over the mark.

(4) The instrument is turned until one of the level bubbles is parallel to the line joining two diagonally opposite leveling screws.

(5) The instrument is then leveled by turning the leveling screws.

Q. How is the recording theodolite set up and leveled? *A.* The instrument is mounted on a permanent pedestal. It is screwed or clamped onto a base ring and is leveled in the same manner as that described for a transit.

Q. What manning detail is required for each recording theodolite? *A.* Two men at each instrument, one tracker and one camera operator.

Q. What is the source of power for the recording theodolites? *A.* Lead-acid storage batteries and dry-cell batteries.

Q. What care is taken in handling observation instruments? *A.* Observation instruments are precision instruments. Avoid jarring. When not in use, keep instruments in carrying cases provided. This will prevent many accidents and will also protect the instruments from dust and moisture. Do not turn the leveling screws so tightly that they bind. Use only optical paper on the exposed optical surfaces.

50. Orientation.—*Q.* How is the AA BC observation instrument M1 oriented? *A.*

(1) Set the main azimuth scale and vernier to read the azimuth of the orienting point.

(2) Loosen the leveling screws, grasp the mount above the leveling screws, and rotate the assembly until the telescope is pointed approximately at the orienting point.

(3) Tighten the leveling screws and level the instrument.

(4) Release the azimuth release clamp and center the telescope on the point.

(5) Tighten the azimuth release clamp.

(6) By means of the tangent screw, center the instrument exactly on the orienting point.

Q. How should the observer's eyepiece be located with respect to the traversing handwheel? *A.* The traversing handwheel should be to the right of the observer. To accomplish this, release the azimuth release clamp and turn the telescope to the proper position. Tighten the azimuth release clamp. This operation must be done before orienting.

Q. How is the flank spotting instrument M1 oriented? *A.*

(1) Set the ϵ_m scale so that the pointer is opposite zero. Lock in position with clamping screw.

(2) Rotate the instrument on the vertical shaft of the base until the pointer "To battery" on the θ scale is directed toward the battery position. Tighten the clamping knob.

(3) Loosen the ϵ_m clamping screw.

Q. How is the recording theodolite oriented? *A.*

(1) The instrument is first leveled. The azimuth handwheel is turned until the counter visible through the glass reads the azimuth of the orienting point.

(2) The clamping ring on the base is loosened and the entire instrument turned around on the base until the cross hair of the operator's telescope bisects the orienting point.

(3) The setting is checked to see that the azimuth reading has not been changed, and the clamping ring is clamped on the base.

51. Adjustment.—*Q.* What adjustments of observation instruments by battery personnel are authorized? *A.* In general, any adjustments which do not require the disassembly of the instrument.

Q. What is collimating? *A.* The process of making the lines of sight through two or more optical instruments on the same mount parallel.

Q. How is collimating accomplished? *A.* In general, by one of two methods.

(1) Using an aiming point at long or infinite range, the lines of sight are laid on the same point using the collimating adjustment. The aiming point might be the sun, some other celestial body, or a well-defined point at such a range that the parallax becomes negligible.

(2) Using an aiming board at close range, marks are made on a vertical surface the same distances apart as the lines of sight. Then each line of sight is made to point at its corresponding marks by means of the collimating adjustment.

Q. Describe the method employed in collimating the two telescopes of the AA BC observation instrument M1, using the aiming board at close range. *A.*

(1) Set up and level the instrument.

(2) On a convenient vertical surface normal to the line of sight, draw a horizontal line at the same height as the trunnions of the instrument. Make two marks on this line separated by the distance between the lines of sight of the two telescopes.

(3) Aline the cross hairs of the battery commander's telescope horizontally and vertically on the right-hand mark.

(4) Using only the collimating adjustments, adjust the cross hairs of the observer's telescope horizontally and vertically on the left-hand mark. The elevation adjustment is accomplished by means of the clamp which holds the observer's telescope to the trunnion. The azimuth adjustment is made by adjusting the spring adjustment at the rear of the observer's telescope.

SECTION II

ORIENTATION AND RECONNAISSANCE

52. Orientation.—*Q.* What is meant by orientation? *A.* As here used it means the determination of length and direction of lines and the location of certain points.

Q. What equipment is used with a transit to measure distances and elevations? *A.* Level rods, stadia rods, steel tapes, and tally pins (or marking chalk).

Q. What is a level rod? *A.* A rod used with a level or transit to determine the vertical distance of the line of sight of the transit above the point on which the rod rests. As elevations are usually measured in feet, the level rod is graduated in feet and tenths.

Q. What is a stadia rod? *A.* A rod used in connection with the stadia wires of the transit to measure horizontal distances. The vertical interval (intercept) on the rod seen between the two horizontal stadia wires in the telescope is read and multiplied by 100 to give the distance from the transit to the rod. Thus a rod intercept of 1.2 feet would correspond to a distance of 120 feet from transit to rod.

Q. How are steel tapes graduated? *A.* Steel tapes are made in various lengths and are graduated in feet or meters; sometimes both, feet on one side, meters on the other.

Q. What care should be taken of steel tapes? *A.* Care should be taken to avoid kinks and the tape should be cleaned and oiled after use.

Q. What are the duties of the rodmen? *A.* The front rodman selects the new station and holds the rod at that point until the instrumentman occupies that station. He then selects a new station forward. The new station must be clearly visible from the instrument and not more than 500 yards away if distances are measured by stadia. The rear rodman occupies the station vacated by the instrumentman, to the rear of the instrument. The rod must be held vertically during sighting.

Q. What are the duties of chainmen? *A.* The rear chainman alines the front chainman on the rod ahead. The front chainman carefully marks off each tape length with tally pins or crayon. Each keeps a record and if they disagree must measure again. Stations should be definitely located on even 10-foot marks of the tape for ease in computations. The tape must be kept *horizontal* with no kinks in it. When taping on steep slopes a plumb bob must be used with the tape.

53. Reconnaissance.—*Q.* What is meant by reconnaissance? *A.* As here used it means the securing of information pertaining to the selection of battery positions, observation posts, command posts, routes of approach, and routes for communications.

Q. What enlisted personnel is included in a reconnaissance party? *A.* Those required and selected to accompany the commander or officer in charge, usually a communication sergeant, an instrument sergeant, and scouts or messengers. Agents of other units may be included in the detail if it is a battalion commander's reconnaissance party.

Q. What are the duties of the communication sergeant? *A.* He assists in selecting communication routes and switchboard stations and makes preparation for establishing them.

Q. What does the instrument sergeant do? *A.* He assists in selecting fire-control stations or observation posts and assists in installing them.

Q. What are the duties of the scouts? *A.* Scouts as members of reconnaissance parties assist in reconnaissance and when spotting OP's are selected they establish them and act as observers and spotters. They may be used also to mark the positions for the guns selected by the battery commander.

Q. What is an agent? *A.* A representative of any unit attached to the headquarters of a higher unit or other command. He must possess training similar to that of a scout. He is frequently detailed to assist in the reconnaissance of a superior commander and is either left at a position selected for his organization or sent back to guide his organization to the position.

Q. What are the qualifications for an agent? *A.* He should be able to read and follow routes from a map, make a simple sketch, drive a motorcycle or car, write a message, deliver an oral message, and understand the requirements sought after in making a reconnaissance. His duties are such that he should be active and intelligent and should have initiative.

Q. What special training should agents and scouts have? *A.* Training in orientation, map reading, identification of targets, and all other duties of an observer or a messenger.

CHAPTER 8

DEFINITIONS

SECTION I

ELEMENTARY DEFINITIONS FOR ANTIAIRCRAFT ARTILLERY

54. General.—The definitions listed in this section are intended
to give a general knowledge of elements of data and terms applicable
to antiaircraft artillery.

Adjustment of fire.—The process of determining and applying corrections to firing data to bring the center of burst, or the cone of
fire, to the adjusting point and to keep it there.

Altitude.—The vertical distance to a point in space from a horizontal
reference plane, usually the horizontal plane containing the directing
point of the battery.

Angular height.—The vertical angle between the line of position
(site) and the horizontal.

Axis of bore.—The center line of the bore of the gun.

Azimuth.—The horizontal angle, measured in a clockwise direction,
from a selected reference line (usually the grid north line) passing
through the position of the observer to the horizonal projection
of the line of sight from the observer to the objective.

Back azimuth.—The azimuth plus or minus 180° or 3.200 mils. The
opposite direction.

Base line.—A line of known length and direction between the primary
(battery) and one of the secondary (flank) observation or spotting
stations, the position of which with respect to the battery is known.
The base line is called *right-handed* or *left-handed* depending on
whether the secondary station is to the right or left of the primary
station from the point of view of a person facing the field of fire.

Base piece.—*See* Directing point.

Bore.—The interior of a gun or cannon forward of the front face of the breechblock (or bolt). The length of the bore is the distance from the front face of the breechblock proper to the muzzle, measured along the axis of the bore.

Center of burst.—The mean position in space of a particular series of bursts.

Dead time.—The time necessary to compute and utilize an element of the firing data.

Deflection.—The angular amount by which the gun must lead the target at the instant of firing in order to hit the target.

Degree.—A unit of angular measure: a circle is divided into 360 equal parts or degrees.

Deviation.—The angular or linear displacement of a point of burst or a center of burst, or the center of a cone of fire, from the target or adjusting point.

Directing point.—A point in or near a battery for which the firing data are computed. If a gun of the battery is the directing point, it is called the base piece or directing gun.

Dispersion.—The scattering of shots fired with the same data.

Displacement.—The distance from one point to another point. Gun displacement is the horizontal distance in yards from the pintle center of the gun to the directing point or directing gun of the battery.

Drift.—The departure of a projectile from the vertical plane in which it is fired, caused by the rotation of the projectile and the resistance of the air. In the United States service, drift is always to the right.

Firing data.—All data necessary for firing a gun at a given objective.

Firing tables.—A collection of data, chiefly in tabular form, intended to furnish the ballistic information necessary for conducting the fire of a particular model of gun with specified ammunition.

Fuze.—A device attached to a projectile which controls the time of burst of the projectile.

Ground speed.—The linear velocity of the target, usually expressed in yards per second, with reference to the ground.

Gunner's quadrant.—An instrument used on the quadrant seat of a cannon to measure the vertical angle between the axis of the bore and the horizontal.

Horizontal range.—The length of the base of the vertical right triangle in space, the vertical side of which is altitude and the hypotenuse of which is the line of position.

Laying.—The operation of pointing a gun in elevation or direction, or in both elevation and direction, without the use of a sight.

Leveling.—The process of adjusting a gun and mount or an instrument so that all vertical or horizontal angles will be measured in true vertical or horizontal planes.

Line of position.—The line of position, or line of site, of a point is the straight line connecting the point of origin with that point. The point of origin is usually a gun or position-finding instrument.

Line of sight.—The line of vision; the optical axis of an observation instrument.

Line of site.—*See* Line of position.

Mil.—A unit of angular measure. One sixty-four-hundredths part of a circle. For practical purposes the arc which subtends a mil at the center of a circle is equal in length to one one-thousandth of the radius.

Muzzle velocity.—The velocity of the projectile at the origin of the trajectory. Also called initial velocity.

Normal.—Geometrically the term means perpendicular to. When used in connection with reference scales, the normal setting is that reference scale setting which corresponds with a true setting of zero.

Orientation.—The establishment of true horizontal lines of known direction. The process of adjusting the azimuth circles of guns or instruments so that they will read correct azimuths when pointed in any direction.

Parallax.—The difference in azimuth or direction of a point as viewed from two other points.

Pintle center.—The vertical axis about which a gun and its carriage are traversed.

Plane of fire.—The vertical plane containing the axis of the bore when the gun is ready to fire.

Plane of position.—The vertical plane containing a line of position.

Pointing.—The operation of giving a piece a designated elevation and direction.

Position finding.—The process of determining the present and future positions of a target for the purpose of directing fire upon it.

Position of target.—Two positions of the target are considered: Present position is the position of the target at the instant of firing; future position is the predicted position of the target at the end of the predicted time of flight.

Predicting.—The process of determining the expected position of the target at some future time.

Quadrant elevation.—The vertical angle between the horizontal and the axis of the bore when the gun is ready to fire.

Reference numbers.—Arbitrary numbers used in place of actual values in the graduation of certain scales. Their purpose is to avoid the use of positive and negative values.

Round.—All of the component parts of ammunition necessary in the firing of one shot.

Sense.—The direction of a point of burst, center of burst, or center of a cone of fire, with respect to the target or other aiming point, as over or short, right or left, above or below.

Slant range.—The hypotenuse of the vertical right triangle in space, the vertical side of which is altitude and the base of which is horizontal range.

Spotting.—The process of determining the position of a point of burst or of the center of a cone of fire with respect to the adjusting point.

Superelevation.—That part of the quadrant elevation which allows for the curvature of the trajectory under the conditions actually existing.

Symbols.—Letters used to represent certain elements of firing data, angles, or position of the target. For example: ϵ represents angular height; φ, quadrant elevation; H, altitude; F, fuze range; and R, horizontal range.

Synchronization.—A process in which the values indicated by all receiver pointers of a data transmission system are made to agree with the values set on the corresponding transmitters.

Time of flight.—The elapsed time from the instant the projectile leaves the bore of the gun to the instant of impact (burst).

Trajectory.—The curve described by the center of gravity of a projectile in flight.

55. Antiaircraft guns.

Angular unit method.—A method of adjusting antiaircraft artillery gunfire in which range deviations in mils obtained by a flank observer are converted into altitude corrections in yards for application at the director.

Directing point.—A point in or near a battery for which the firing data are computed. If a gun of the battery is the directing point, it is called the directing gun.

E–W and N–S rates.—The continuous tracking of the target establishes an instantaneous rate of speed (S_g) relative to the ground. This rate is resolved into an east-west component (E–W rate) and north-south component (N–S rate).

E–W and N–S travel (ΔX *and* ΔY).—The E–W and N–S rates multiplied by time of flight give E–W and N–S travel during the time of flight of the projectile.

Fuze range.—The fuze setting necessary to produce a burst at a given point in space.

Gun difference.—The difference, due to displacement, between the range from a gun to the target and the range from the directing point to the target.

Trial fire.—Deliberate fire at a point in space to determine corrections for firing data. Normally five rounds are fired, all from the same gun.

Trial shot point.—A point in space at which trial fire is conducted.

X_o *and* Y_o.—The horizontal range to the present position of the target is resolved into components in east-west direction (X_o) and north-south direction (Y_o).

X_p *and* Y_p.—The horizontal range to the future (predicted) position of the target is resolved into components in the east-west direction (X_p) and north-south direction (Y_p).

56. Antiaircraft automatic weapons.

Central control.—A method of fire control for automatic weapons in which the leads are controlled from a central point rather than by the individual gunner.

Individual control.—A method of fire control for automatic weapons in which the leads are controlled by the individual gunner.

Initial lead.—A lateral or vertical lead applied to the gun sights before firing is commenced.

Lateral lead.—The angle in the slant plane of the lateral sight by which the gun must lead the target to cause the projectile and target to meet.

Vertical lead.—The angle by which the gun must lead the target vertically in order that the projectile will meet the target at the future position. It is measured in the vertical plane containing the axis of the bore of the gun.

57. Antiaircraft searchlights.

Acoustic corrections.—Corrections to sound locator data for nonstandard atmospheric conditions and sound lag.

Aerial sound ranging.—The process of locating aircraft by means of the sounds emitted.

Arbitrary corrections.—Corrections to sound locator data which are applied to correct for observed errors after all known deviating causes have been corrected for.

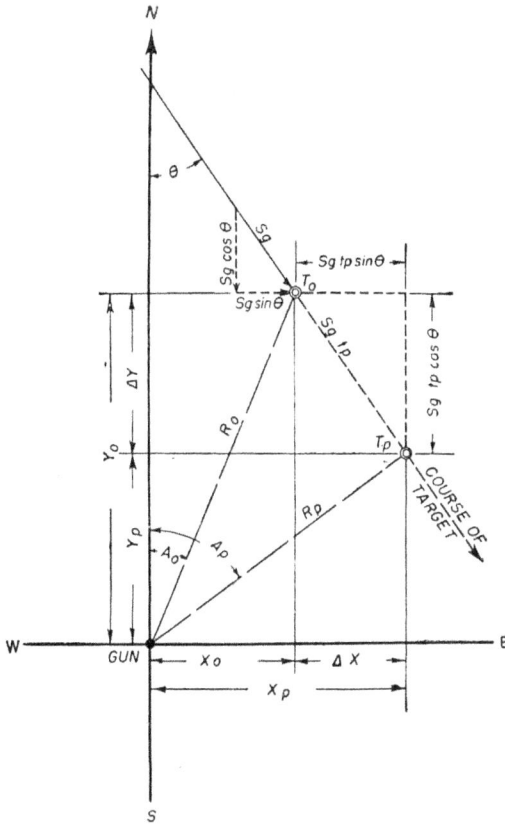

FIGURE 79.—Elements of data, linear speed method (horizontal projection).

A_o	Azimuth of target at present position (T_o).
A_p	Azimuth of target at future position (T_p).
ΔX or $S_g t_p \sin \theta$	} East-west component of travel of target during time of flight of projectile.
ΔY or $S_g t_p \cos \theta$	} North-south component of travel of target during time of flight of projectile.
R_o	Horizontal range to target at present position (T_o).
R_p	Horizontal range to target at future position (T_p).
S_g	Ground speed of target.
$S_g \cos \theta$ or N-S rate	} North-south component of ground speed of target.
$S_g \sin \theta$ or E-W rate	} East-west component of ground speed of target.
$S_g t_p$	Linear travel of target in horizontal plane during time of flight.
θ	Angle between vertical planes containing course of target and north-south axis of director (never greater than 90°).
T_o	Position of target at instant of firing (present position).
T_p	Future or predicted position of target.
t_p	Time of flight to future position of target (T_p).
X_o	East-west component of horizontal range to present position (T_o).
X_p	East-west component of horizontal range to future position (T_p).
Y_o	North-south component of horizontal range to present position (T_o).
Y_p	North-south component of horizontal range to future position (T_p).

179

Distant electric control.—A system for the control of the pointing of searchlights from a distance. The system consists of the controller at the control station and the necessary motors or receivers at the searchlight.

Sound lag.—The angular difference between the actual (present) position of the target and the apparent position as indicated by sound.

Zero reader.—A device for indicating when the searchlight is properly pointed on corrected sound locator data.

SECTION II

PARTICULAR DEFINITIONS PERTAINING TO SUPPLIES (INCLUDING AMMUNITION) AND SUPPLY FUNCTIONS

58. General.

Credit.—An allocation of a definite quantity of supplies which is placed at the disposal of the commander of an organization for a prescribed period of time.

Distribution point.—A place, other than a depot or railhead, where supplies are issued to regiments and smaller units. Distributing points are designated by the class of supplies therein, and by the identity of the unit establishing them; such as "Class I Distributing Point, 1st Division," or "Ammunition Distributing Point, 1st Infantry."

Dump.—A temporary stockage of supplies established by a corps, division, or smaller unit. When supplies are ordered issued from dumps, the latter become distributing points. Dumps are designated by the identity of the unit establishing them and by the class of supplies therein, such as "1st Infantry Ammunition Dump" or "1st Division Class 1 Supply Dump."

Railhead.—A supply point on a railroad where loads are transferred from rail transportation to some other type of transportation. Railheads are designated in the same manner as distributing points; for example, "Class I Railhead, 1st Division," or "Ammunition Railhead, 1st and 2d Divisions."

Requisitions.—Requests for supplies, normally submitted on the prescribed form, to a higher commander. When approved by the higher commander, a requisition becomes an order for issue of supplies by the proper supply agency to the supply officer of the unit which submitted the requisition.

Shipping ticket.—A form which accompanies a shipment of supplies to a supply officer and which he must sign and return to the shipping officer to accomplish transfer of accountability.

Supply point.—A general term used to include depots, railheads, dumps, and distributing points.

Train.—That portion of a unit's transportation, including personnel, operating under the immediate orders of the unit commander primarily for supply, evacuation, and maintenance. It is designated by the name of the unit; such as "1st Infantry Train."

59. Ammunition.

Ammunition train.—Transportation and personnel of the battalion organized to supply ammunition to the batteries and to carry a reserve supply of ammunition.

Periodic ammunition (expenditure) reports.—Reports of the expenditure of ammunition by the unit concerned, normally covering the 24-hour period immediately preceding the time of rendition of the report. These ammunition reports are the basis for the establishment of additional ammunition credits for the unit concerned.

Unit of fire (formerly called "day of fire").—The quantity in rounds or tons of ammunition, bombs, grenades, and pyrotechnics which a designated organization or weapon may be expected to expend on the average in 1 day of combat.

60. Supplies other than ammunition.

Automatic supply.—A process of supply under which deliveries of specific kinds and quantities of supplies are moved in accordance with a predetermined schedule. *Daily automatic supply* means that supplies are dispatched daily to an organization or installation.

Classes of supplies.—There are five classes of supplies:

Class I.—A class of supplies consisting of those articles which are consumed at an approximately uniform daily rate irrespective of combat operations or terrain, and which do not necessitate special adaptation to meet individual requirements; such as rations and forage.

Class II.—A class of supplies consisting of those authorized articles for which allowances are established by Tables of Basic Allowances; such as clothing, gas masks, arms, trucks, radio sets, tools, and instruments.

Class III.—A class of supplies consisting of engine fuels and lubricants, including gasoline for all vehicles and aircraft, Diesel oil, fuel oil, and coal.

Class IV.—A class of supplies consisting of those articles which are not covered in Tables of Basic Allowances and the demands for which are directly related to the operations contemplated or in progress (except for articles in classes III and V) ; such as fortification materials, construction materials, and machinery.

Class V.—A class of supplies consisting of ammunition, pyrotechnics, antitank mines, and chemicals.

Forage.—Food for animals. To collect supplies for men and animals.

Issue.—A delivery of supplies. Specifically, the delivery of supplies of any kind by a supply department to responsible persons authorized to receive them on behalf of their organizations. Also the supplies so delivered.

Memorandum receipt.—A receipt given to a supply officer by a person drawing supplies from him, or a receipt given by the supply officer to a person returning supplies to him.

Railhead distribution.—Issue of class I supplies to regimental (or similar unit) transportation at the railhead.

Ration.—The prescribed allowance of the different articles of food for the subsistence of one person or one animal for 1 day.

CHAPTER 9

MOTOR TRANSPORTATION

61. Nomenclature of major parts of motor vehicles.—*Q*. Into what general groups may the parts of any motor vehicle be divided? *A*. Power plant, transmission system, control system, chassis, running gear, and body.

Q. Point out the principal parts of the power plant to include the crankcase, cylinders, valves, and various parts pertaining to fuel, carburetion, ignition, lubrication, and cooling system.

Q. Point out the principal parts of the transmission system to include the clutch, transmission, driveshaft, universal joints, differential, torque arms, and axles.

Q. Point out the principal parts of the control system, chassis, and running gear to include: frame, springs, brake drums, brake rods, wheels, steering knuckle, drag link arm, and emergency and foot brakes.

Q. Why are instruments installed on the dash? *A*. For the purpose of indicating and controlling the operation of the engine and vehicle.

Q. What instruments are usually installed? *A*. Ammeter, oil pressure gage, speedometer, thermometer (engine temperature), choke, light switch, hand throttle, spark control, and ignition switch.

Q. What does the ammeter indicate? *A*. The amount of current that is being consumed by the ignition and light system of the vehicle from the battery, or the amount of current that the generator is supplying to the battery.

Q. What does the oil gage indicate? *A*. Oil pressure only. The quantity of oil is indicated by the dip stick. Lack of oil or oil pressure is very serious.

Q. Does the oil pressure gage indicate that the engine is being lubricated? *A*. No; it only indicates that the pump is forcing oil some place at the pressure indicated by the gage.

62. Practical operation of motor vehicles, to include driving and fueling.—*Q.* What are the qualifications for a good driver? *A.* Good physical condition and common sense.

Q. Define common sense in connection with driving. *A.* Alertness, judgment, and caution on the road. A good driver will obey

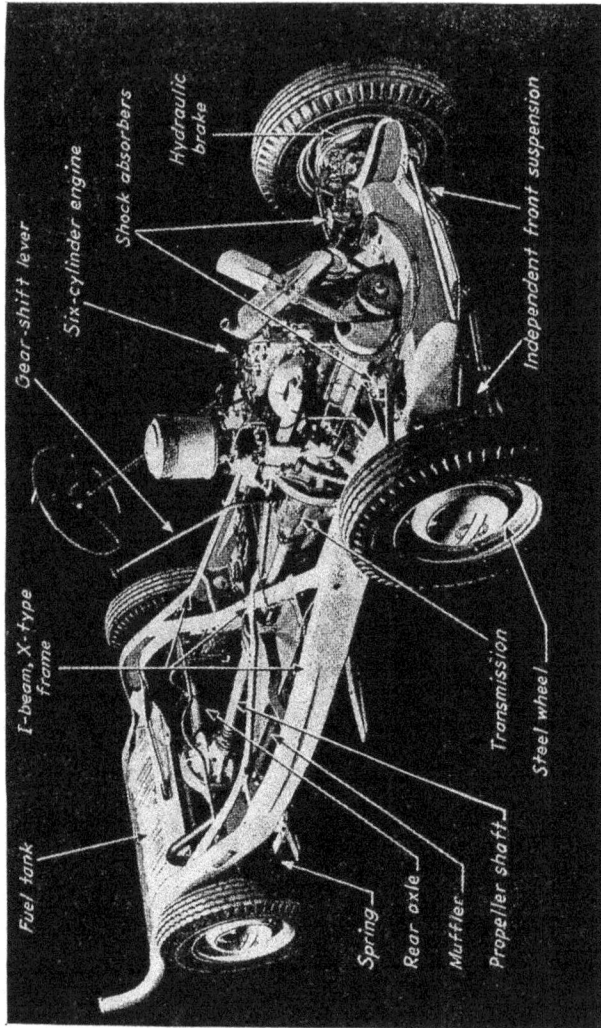

FIGURE 80.—Modern gasoline engine-driven passenger car chassis showing its various parts.

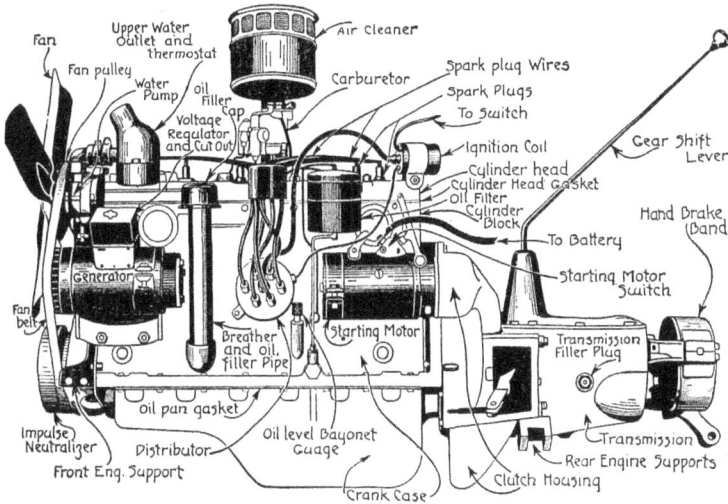

FIGURE 81.—Typical engine and transmission assembly showing external parts, equipment, and accessories.

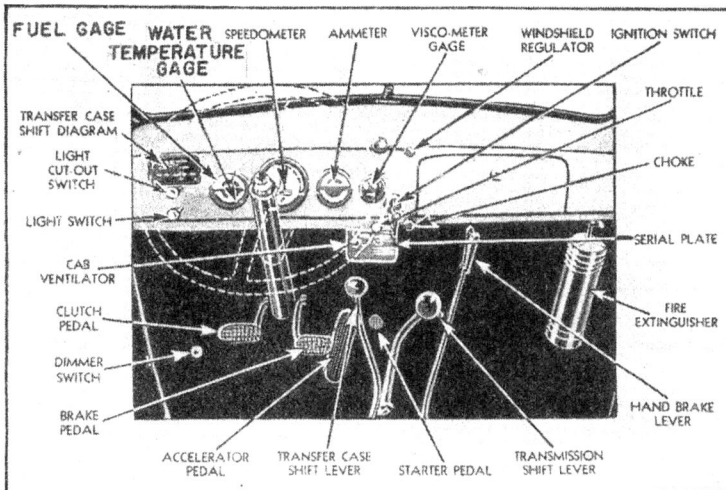

FIGURE 82.—Interior view of truck cab showing instruments and controls.

all traffic regulations and carry out the rules of good maintenance driving. He will respect the rights of other drivers and of pedestrians.

Q. Mention the most important rules to be observed on the road.

A. (1) Have vehicle under control at all times.

(2) Never exceed prescribed speed limits nor the speed limit of your vehicle.

(3) Keep a safe distance in rear of a vehicle in front so you can stop if that vehicle stops suddenly.

(4) Keep on the right side of the road.

(5) Do not try to pass a car parked or moving on your side of the road if a car is approaching from the opposite direction, except when operating on a road having three or more lanes.

(6) Do not try to pass a car on a hill or curve unless you can see the road far enough ahead to assure yourself that no car is coming in the opposite direction.

(7) Sound the horn before passing a car going in the same direction.

(8) Give the proper hand signal before stopping or turning.

(9) Go slowly on sharp curves.

(10) Do not pass street cars taking on or discharging passengers except where safety zones are provided.

(11) Slow down when roads are slippery.

Q. List a few rules whose observance will help to prevent accidents.

A. (1) Obey all traffic regulations and special instructions. This includes using the proper hand and horn signals.

(2) Never depend on what the other operator or pedestrian may do.

(3) Never operate a vehicle with faulty brakes, steering mechanisms, or lights.

Q. What precaution is taken before stopping or turning a corner?

A. Signal to drivers of other vehicles, extending arm in the proper signal. Before turning corners or sharp curves, slow down, sound the horn, and be prepared to stop to avoid collisions with other cars which may be hidden from view. The same precautions are taken at street intersections or crossroads which are not clearly visible for some distance in each direction.

Q. What inspections are required to be made by the driver before leaving and after returning to the garage?   *A.*

(1) Oil level in crankcase.

(2) Water in radiator.

(3) Gasoline supply.

(4) Condition of tires and battery.

(5) Inspection for leaks in cooling and oiling systems.

(6) Mechanical condition of vehicle, especially brakes, steering, lights, and horn. Any faults and unusual noise observed during operation should be reported to dispatcher immediately.

Q. Give several common faults in driving which are damaging to the vehicle. *A.*

(1) Racing the engine at any time.

(2) Inadequate use of the gears.

(3) Excessive or improper appliance of the brakes.

(4) Riding, slipping, or quickly engaging the clutch.

(5) Turning front wheel, while standing.

(6) Excessive use of the choke.

(7) Excessive speed in first or second gears.

(8) Continuing to drive with minor maladjustments.

Q. What precautions should the driver take against fire? *A.*

(1) Never refuel while engine is running.

(2) Be careful not to let the fuel tank overflow, as hot exhaust pipe and manifold can readily ignite the fuel.

(3) Keep old oily rags, waste, and papers from under the seat.

(4) Keep engine clean.

(5) Do not smoke while driving or on the vehicle.

Q. What should be done in case a vehicle catches fire? *A.* If the vehicle is inside a building, push it out, if possible. Use the fire extinguisher that is carried on every Government vehicle, playing it directly on source of fire. Do not use water on a gasoline or oil fire; it only tends to spread it. If the fire extinguisher is not sufficient to extinguish the fire, use dirt, sand, or mud; in some cases it can be smothered by using such articles of clothing as may be available. If fire should break out in the load of the vehicle, remove load until the source of the fire can be reached.

Q. What precautions are necessary in cold weather? *A.*

(1) Protect the water in the radiator from freezing.

(2) Watch condition of battery, as it does more work and is less efficient in cold weather.

(3) Use chains or tractioneers when necessary.

(4) Keep off soft or partially frozen ground.

Q. How should radiator be protected in cold weather? *A.* Unless filled with antifreeze solution, the radiator and water jackets should be completely drained, when the vehicle is not in use, and a "drain" sign hung on the radiator. Sometimes it will be necessary to protect the lower front-half of the radiator with tin, cardboard, or canvas.

Q. What precaution should be observed when filling the radiator when engine is very hot? *A.* The engine should be stopped and allowed to cool off before adding water. If time will not allow this, let the engine run while water is added slowly (preferably warm water).

Q. What is the proper way of applying the foot or service brake? *A.* It must be applied, except in case of emergency, with evenly increasing pressure; as the vehicle comes to a stop the pressure should be progressively reduced to give a smooth stop. Sudden stops are hard on the vehicle and on the brakes and may cause rear end collisions.

Q. How should air brakes be applied? *A.* The best possible stop will be made when the brakes are applied at the very start as hard as the speed and condition of the road will permit, and then eased off as the speed is reduced, so that at the end of the stop but little pressure remains in the brake chambers. In easing the brakes off, do not "fan" the brake valve, repeatedly releasing and applying the brakes, as this wastes air pressure.

Q. What inspection of air brakes should be made before starting the vehicle? *A.* Observe the air pressure gage, showing the pressure stored in the reservoir. It must read 40 pounds or over before the air brakes can develop full effectiveness.

Q. What is the purpose of the hand brake? *A.* To hold the vehicle in a parked position. In emergency to relieve the foot brake. Caution must be exercised in applying the hand brake because if it is of the propeller shaft type a sudden application may strip the rear end gears and the vehicle will be out of control.

Q. What is the purpose of the choke? *A.* To restrict the air passage at the inlet of the carburetor, thereby giving a rich mixture for starting and warming up the engine.

Q. What is the proper use of the choke? *A.* To assist in starting when the engine is cold or the vehicle has been left idle for some time. Excessive use will flood the engine, making starting temporarily impossible and interfering with proper lubrication.

Q. What throttle setting should be used for starting? *A.* This depends upon the vehicle. Most carburetors are designed so that the proper setting for starting is determined by a throttle stop. By stepping on the accelerator a few times before starting, the engine will be primed, and the engine should start when the starter is engaged.

Q. What throttle setting should be used until the engine warms up? *A.* A setting corresponding to about 20 mph vehicle speed. The engine should not carry a load during this period.

Q. What is the proper use of the accelerator? *A.* The accelerator should be depressed slowly. Tramping on the accelerator floods the engine, wastes gasoline, and fouls the spark plugs.

Q. What precautions must be taken with the ignition switch? *A.* It must be left locked, whenever the vehicle is parked, to prevent damage to the coil and battery.

Q. How do you select the proper gear? *A.* A gear is selected that will allow the engine to run without lugging. If the engine cannot reach its governed speed, gears should be shifted. When descending grades a gear must be selected that will not force the engine to run faster than its governed speed.

Q. What damage might result from the improper selection of gears? *A.*

(1) Engine bearings might be damaged or cylinder head gasket blown.

(2) The driver might be forced to shift down two gears and thus delay the convoy.

(3) If on a downgrade the truck might run away or the engine may turn up so fast that it will be damaged.

Q. What is the proper way to shift gears? *A.* Bring the engine to full governed speed in each gear as the shift is accomplished.

Q. Explain double-clutching and its purpose. *A.* Double-clutching is accomplished by engaging the clutch while the transmission hesitates in neutral when the gears are being shifted up or down, then shifting to the next gear in the normal manner. During the hesitation period the foot is removed from the throttle if the shift is from a lower to a higher gear; if from a higher to a lower gear the engine is speeded up to the speed that it should be running in the lower gear selected. Double-clutching is useful in shifting from a lower to a higher gear on trucks that are hard to shift. It is useful in changing to a lower gear preparatory to descending a grade. Double-clutching has its limitations, and the driver must not wait until it is too late to shift. (All drivers should be required to shift from a higher to a lower gear without clashing gears.)

Q. Can front-wheel drive clutches be used while the vehicle is in motion? *A.* Yes. Levers must not be forced. In some instances the front wheels may have to be turned to allow shifting dogs to come into line.

(Drivers of all vehicles should be required to master these special shifting devices.)

Q. What is an auxiliary transmission? *A.* An over-, under-, and direct-drive gearing used in conjunction with the transmission.

Q. How is an auxiliary transmission operated? *A.* When operating under ordinary road and load conditions, it is placed in direct-drive position. When operating under difficult road conditions, or over uneven roads or steep grades with capacity loads, it is placed in the under-drive position. When operating over level roads with light loads, it may be placed in the over-drive position to give maximum road speed without excessive engine speed.

Caution: The auxiliary transmission must never be shifted while vehicle is in motion.

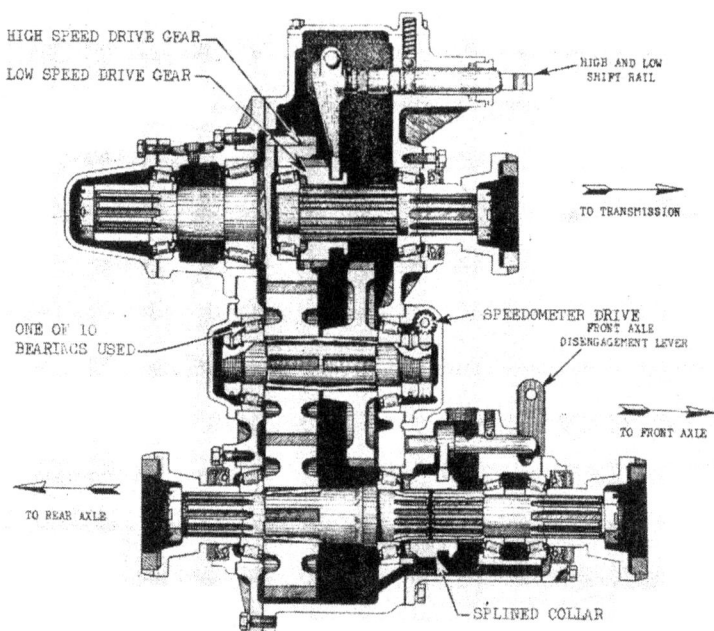

FIGURE 83. –Typical two-speed transfer case showing relation to driving and driven units and disengaging feature for front axle.

Q. How is a car brought back to the center of the road after beginning to skid on a wet pavement or muddy road? *A.* When the rear of the car starts to skid, turn the steering wheel in the direction the car is skidding and partially close the throttle. To close the throttle entirely would have the same effect as applying the brakes. Do not apply the brakes. When skidding on a narrow road, it is best to apply more power and steer for the center of the road. This will aggravate the skid for a moment but will bring the car around at an angle with the front wheels in the center of the road. The momentum of the car will cause the rear wheels to climb back onto the road.

Q. What is the normal operating temperature of a gasoline engine? *A.* Approximately 180° F.

Q. Where is this temperature taken? *A.* In the water that surrounds the cylinders and combustion chambers of the engine.

Q. What data are usually found on the dash plate? *A.* Make and model of vehicle, maximum speed, tonnage that the vehicle was designed to carry, engine number, serial number, and date of manufacture.

Q. How should a vehicle be loaded? *A.* The load should be distributed equally, fore and aft, and to the right and left, of the center of gravity of the vehicle. Heavy items are placed on the bottom of the load. It should be systematically loaded to facilitate delivery. Any load beyond the capacity of the vehicle should be refused. The load should be properly secured by lashing or some other means. Red flags or lanterns must be attached to all loads protruding beyond the truck body.

Q. How can the driver tell if the vehicle is overloaded? *A.* By noting the set of the springs. The candidate will be required to—

(1) Start the engine of a truck or car.

(2) Start in first gear, shift into second and third gears.

(3) Shift back into second gear.

(4) Stop the vehicle.

(5) Shift into reverse gear, and back the vehicle; and

(6) Shift into neutral and stop engine.

63. Trouble shooting and minor repairs.—*Q.* What is the most usual cause of engine trouble? *A.* Ignition. In the field, dirt and water in fuel run ignition trouble a close second.

Q. Before making a detailed investigation of engine trouble what tests should be made? *A.*

(1) That clean gasoline is reaching the cylinders. If in doubt the cylinders should be primed.

(2) That the spark is occurring and all wires are attached.

(3) That compression is satisfactory as tested with the crank.

(4) That ignition timing is approximately correct.

Q. How can a check be made that gasoline is reaching the cylinders? *A.* Disconnect the gasoline line at the carburetor, turn the engine over, and see if gasoline is pumped from the line.

Q. How can it be determined if a spark is occurring? *A.* Turn on ignition, remove one of the spark plug wires, and hold it by its insula-

tion a short distance from the engine; turn the engine over and note
if spark occurs.

Q. How can it be determined if ignition timing is approximately
correct? *A.* Remove a spark plug but leave its wire attached and remove valve cover. Through spark plug hole note when piston is at
top dead center and at the same time note when both valves are closed.
Continue to turn engine over slowly and note when spark occurs. To
be correct, spark should occur at or near top dead center with both
valves closed.

Q. If the four basic tests are positive but the engine still refuses to
start, what may be the trouble? *A.*

(1) Engine flooded.
(2) Choke not working. .
(3) Carburetor frozen.
(4) Engine too cold.
(5) Valve sticking open.
(6) Valve spring broken.
(7) Spark plugs dirty or with too wide a gap.
(8) Poor gasoline.
(9) Wiring out of order.
(10) Wet ignition system.
(11) Battery too weak.
(12) Throttle levers disconnected.
(13) Carburetor jet plugged.
(14) Condenser weak.
(15) Oil too heavy.
(16) Blocked muffler.

Q. What does light blue smoke from the muffler indicate? *A.*
Burning of oil for some reason.

Q. What does steam from the exhaust indicate? *A.* A water leak
due to a blown gasket or cracked engine.

Q. What does black smoke from the exhaust indicate? *A.* Too
rich a mixture. Engine will be sluggish when this condition exists.

Q. How does the driver generally locate trouble? *A.* By inspections, generally during operation.

Q. What repair parts should the driver carry? *A.* Tape, wire, tire
patching outfit, extra spark plug, and such extra parts as past expe-
rience has shown are liable to frequent failure.

Q. What should the driver do when his vehicle is being repaired?
A. He should assist the mechanic and point out past troubles.

Q. Before performing any repairs what should be done? *A.* The
motor vehicle instruction book should be consulted.

Q. What are indications of steering trouble? *A.*

(1) Play or rattle in a steering gear.

(2) Shimmy.

(3) Peculiar or rapid tire wear.

(4) Hard steering.

Q. What does backfiring indicate? *A.*

(1) A lean mixture.

(2) Carburetor or fuel trouble.

(3) Overheating of engine.

(4) Stuck valves.

(5) Retarded spark.

Q. Name some clutch troubles which should be reported. *A.* Slipping, grabbing, noisy clutch, clutch that will not release.

Q. When do brakes need adjustment or repair? *A.* When they will not stop the vehicle within 30 feet from 20 mph on a dry, smooth, level road.

Q. When should repairs be made to a vehicle? *A.* As soon as they can be done competently.

Q. What is needed to find trouble on a motor vehicle? *A.* A set of testing equipment such as is furnished for the use of each battery.

64. Duties of driver in care, service, repair, and maintenance of motor vehicles.—*Q.* What defines the duties of the driver? *A.* FM 25–10, Technical Manuals of the 10 series, AR 850–15, Circulars 1–10, OQMG, and the motor vehicle manual issued with each vehicle.

Q. How are drivers selected? *A.* On the basis of their standing in an examination on the course of instruction laid down in FM 25–10.

Q. What are the responsibilities of the driver? *A.*

(1) Operation and maintenance of motor vehicles in accordance with instructions.

(2) Care and condition of vehicle, tools, and equipment.

(3) Loads and loading.

(4) Reports and records.

Q. With what should the driver be thoroughly familiar? *A.*

(1) Fire precautions and fire fighting methods.

(2) Accident prevention.

(3) Purpose of the major units of the motor vehicle.

(4) Motor vehicle controls.

(5) Inspections.

(6) Maintenance.

(7) That part of the motor vehicle manual that pertains to the driver.

Q. What may be used to fight a gasoline fire? *A.* Sand or a special extinguisher of the foam, CO_2, or carbon tetrachloride type. *Never use water.*

Q. How is the vehicle fire extinguisher used? *A.* By removing it from its bracket, unlocking it, and pumping. The stream of liquid must be directed at the top or to the windward side of the flame since the liquid releases a gas which is heavier than air.

Caution: This gas is harmful to breathe.

Q. Where are motor vehicle keys kept when the vehicle is in the garage? *A.* They are kept in the vehicles or on a plainly marked board nearby so that vehicles may be moved quickly in case of fire.

Q. May the driver remove gasoline from his fuel tank? *A.* No. The regulations forbid the use of gasoline for cleaning purposes. Gasoline for all authorized purposes may be obtained on a regular issue slip.

Q. For what accidents are drivers responsible? *A.* All accidents that occur to their vehicles while in motion, when parked in an unauthorized place, or when being worked on by themselves.

Q. Define first echelon maintenance. *A.* First echelon maintenance includes all the maintenance functions required to be performed by the driver and his assistant, using only the tools and spare parts on his truck. It is divided into three parts: inspection, preventive maintenance, and repairs.

Q. What inspections are required to be made by the driver? *A.*
(1) *During operation.*—(*a*) Note abnormal readings of dash gages.
(*b*) Note unusual engine sounds.
(2) *At the halt.*—(*a*) Check for fuel, oil, and water leaks.
(*b*) Check tires, tracks, and traction devices.
(*c*) Check for overheating of mechanical units such as brake bands, transmission, etc.
(*d*) Check lights, horn, windshield wiper, etc.
(*e*) Inspect cargo.
(3) *After operation.*—(*a*) Check all items noted in (2) above.
(*b*) Check for loose parts or linkages.
(*c*) Check tools and equipment.
(4) Report results of inspection in each case to the truckmaster.

Q. For what type of maintenance is the driver responsible? *A.* Scheduled, operating, and precautionary maintenance.

Q. What is scheduled maintenance? *A.* Cleaning, lubrication (except when done by a service department), tire care, battery care (except when done by a battery expert), minor repairs, checking of fuel, air, oil, antifreeze, and water.

Q. What constitutes operating maintenance? *A.* Loading. speed, proper use of controls. emergency repairs.

Q. What constitutes precautionary maintenance? *A.* Minor repairs performed as the result of inspections.

Q. What repairs and adjustments may the driver make? *A.* Except for repairing tires and emergency roadside repairs the driver is not permitted to make any repairs or adjustments except under the supervision of the truckmaster.

Q. What maintenance must be performed by the driver on an air-brake system? *A.* Drain the reservoir. daily in cold weather and weekly in warm weather. by opening the drain cock on the bottom. This allows any water collected in the reservoir to run out. *Be sure to close the drain cock after the water has been removed.*

Q. How is the proper spark setting determined? *A.* If the engine runs with full power without knocking, the spark setting is satisfactory.

Q. What care must be taken of the clutch? *A.*

(1) The clutch must not be slipped: gears should be used instead.

(2) When the clutch needs adjustment a prompt report should be made to the motor sergeant.

(3) The clutch must not be let out suddenly. or damage to the whole vehicle will result.

(4) The clutch must be properly lubricated. but must not be over-lubricated or it will slip.

Q. What general precautions should be taken by the driver when working on his vehicle? *A.*

(1) He should not start the engine unless the controls are in neutral.

(2) When working under a truck. he should not depend upon jacks but the vehicle should be firmly blocked.

(3) To lessen the danger of fire he should remove the battery in case of doubt or major repair.

(4) He should work in a well-ventilated place. (During examination the candidate should be required not only to explain the following motor vehicle units and controls but also to demonstrate their proper use to prevent damage to the motor vehicle.)

Q. How does the driver get needed repairs done to his vehicle? *A.* He turns in a bad order report to his truckmaster. The report may be either written or oral.

Q. What records must the driver keep? *A.* Accident report. trip ticket, bad order report. and in some cases an issue slip and vehicle log. These records are kept as directed by the truckmaster.

Q. What kind of oil should be added to the crankcase? *A.* The kind recommended in the instruction manual.

65. Convoy and march rules and discipline.—*Q.* What is a convoy? *A.* A group of two or more military motor vehicles moving as a unit under competent military authority.

Q. What is the purpose of a convoy? *A.* The efficient transportation of personnel and material especially with respect to time required and condition upon arrival.

Q. What is the assigned minimum distance for trucks in convoy? *A.* (1) *Open formation.*—100 yards.

(2) *Closed formation.*—Twice the speedometer reading in yards.

(3) *At halt.*—2 yards.

(4) *Between sections.*—3 to 5 minutes driving time.

Q. Describe drivers' arm signals. *A.*

(1) *Turn right.*—Extend the left arm outward at an angle of 45° above the horizontal.

(2) *Turn left.*—Extend the left arm outward horizontally.

(3) *Slow or stop.*—Extend the left arm outward to an angle of 45° below the horizontal.

(4) *Pass and keep going.*—Extend the left arm horizontally and describe small circles toward the front with the hand.

Q. Describe the commands and signals commonly used in a motorized unit. *A.*

(1) *Start engine.*—Simulate cranking.

(2) *Ready to start.*—Senior in truck stands on running board, faces leader, and extends arm vertically, fingers extended and joined, palm toward the leader.

(3) *Stop engines.*—Cross arms in front of body at the waist and then move them sharply to the side. Repeat several times.

(4) *Increase speed.*—Carry closed fist to the shoulder and rapidly thrust it vertically upward several times to the full extent of the arm.

(5) *Close up.*—Extend the arms horizontally straight to the front, palms in. Move the hands together and then resume the first position. Repeat several times.

(6) *Open up.*—Extend the arms horizontally straight to the front, palms out. Move the hands outward and then resume the first position. Repeat several times.

(7) *Danger.*—Use three long blasts of a whistle or automobile horn repeated several times or three equally spaced shots with a rifle or pistol. The person giving the signal points in the direction of impending danger. This signal is reserved for warning of air or mechanized attack, or other immediate and grave danger. Other signals may be found in FM 25–10.

Q. What are the driver's principal duties during a convoy?   *A*.

(1) Attention to orders and to his driving.

(2) Constant inspections before, during, and after operation.

Q. What should the driver do if he has any trouble when the convoy is on the march?   *A*.

(1) If it is a major trouble he should pull to the side of the road and signal the following vehicle to pass. He should then report his trouble to the maintenance officer, who is at the rear of the convoy. If left behind, the driver will remain with his truck as a guard.

(2) If it is a minor trouble, he will report it to the section mechanic or maintenance officer at the next halt.

Q. What should the driver do during halts of a convoy?   *A*.

(1) He should make the inspections prescribed.

(2) He should keep to the right of his vehicle.

Q. What are the duties of the assistant driver during a convoy?   *A*.

(1) He assists the driver in backing, parking, etc.

(2) He watches to the rear.

(3) He takes his turn at driving.

(4) He assists in first echelon maintenance.

Q. How is gasoline obtained on convoy?   *A*.

(1) In an emergency, from 10-gallon cans carried with the convoy.

(2) At halts, from tankers or some type of filling station.

66. Handling of trucks under adverse conditions.—*Q*. What equipment is furnished each truck and tractor for this purpose? *A*. One tool set (complete with tools) and pioneer equipment motor vehicle set No 1. This set consists of a shovel, pick mattock, an axe, and a bracket to carry them with. One set of chains and in some cases traction devices are also furnished.

Q. What other equipment is available?   *A*. The maintenance section has a block and tackle set, a wrecking set, towbars, and rope. Some vehicles are equipped with power-driven winches, and all vehicles are equipped with towhooks and pintles.

Q. In applying chains what precautions must be taken?   *A*. They must be adjusted properly. In the case of all-wheel drive vehicles they must be placed on all wheels, or broken axles will result.

Q. In case your vehicle gets stalled, what do you do?   *A*. I, or the assistant driver, investigate the reason for stalling and make a plan as to how best to get out of the position. If my decision, or the decision of some one in authority, is that a wrecker is needed, I await the wrecker.

Q. What four abilities must a motor vehicle have to get out of or keep going in a difficult situation?   *A*.

(1) *Power.*—All new trucks issued to the service have enough power.

(2) *Momentum.*—This depends on the speed of the vehicle. In some cases too much speed causes the vehicle to lose part of its traction, resulting in spinning of the wheels. If this occurs the vehicle may become badly stalled.

(3) *Traction.*—All of the multiwheel vehicles are designed to give great traction.

(4) *Flotation.*—This is the ability of the vehicle to ride the ground surface.

Q. How should a difficult hill be negotiated? *A.* On approaching, a sufficiently low gear should be selected to negotiate the hill, and maximum practicable momentum should be obtained. If in column do not start up the hill until the truck ahead has negotiated it. In case of failure, back down in gear. **Caution:** Check to see if brakes hold before shifting to reverse gear, and take steps to enable the truck to climb the grade; for example, lower gear, use of traction devices, or tow from a tractor.

Q. What is a prolonge? *A.* It is a rope with a hook or loop on one end used to maneuver a vehicle by manpower. Prolonges are usually used in pairs.

Q. How should very steep, dangerous slopes be descended? *A.* Straight down, with all personnel except the driver dismounted. Gears should be used, and if brakes are also needed care must be exercised to prevent locking of the wheels. The ignition should not be turned off. Outside assistance may be needed; for example, block and tackle, winch, or prolonges.

Q. What is the best way to negotiate mud? *A.*

(1) Maintain momentum.

(2) Use highest gear possible.

(3) Apply power gently to prevent wheel slippage.

(4) Use traction devices.

(5) Use care in selecting track.

Q. In case a vehicle becomes stalled in mud what should be done? *A.*

(1) If loaded with personnel, have them dismount and push. Sometimes backing up and selecting another way out will solve the problem.

(2) Use a tow. **Caution:** Because of the danger of slipping under the vehicle, personnel should be cautioned against pushing on the side of a moving vehicle that has slipped into a ditch or old wheel ruts.

Q. In case of operating alone what is done if the vehicle becomes stalled? *A.*

(1) Traction may be improved by means of wheel mats, brush, or boards.

(2) The truck may be dug out.

(3) If the vehicle has dual wheels a rope may be used between the wheels: the truck will wind the rope up like a windlass.

(4) A pole may be inserted between the wheels that are slipping. This method is very effective on track-laying vehicles.

Q. What is the most useful device furnished the coast artillery for negotiating difficult terrain? *A.* The power-driven winch on tractors and gun trucks.

Q. How is sand negotiated? *A.*

(1) By the use of traction devices.

(2) By using the same track.

(3) By making roads from chicken wire or brush.

Q. In case skidding occurs what should be done? *A.* The accelerator should be released gradually and the front wheels turned in the same direction that the rear wheels are skidding. Where necessary, prolonges may be used to prevent skidding in very slippery places.

Q. How is a narrow ditch crossed? *A.*

(1) Small ditches, less than the diameter of the tire, or wider shallow ditches should be crossed at an angle. Since this puts a strain on the vehicle the load should be lightened if possible, and personnel should assist at the critical point.

(2) Wide ditches must be filled or bridged before crossing. They are crossed at right angles.

Q. How are shallow streams forded? *A.* Slowly in a low gear.

Q. What precautions should be observed in crossing bridges? *A.*

(1) The speed and load signs should be observed.

(2) When the capacity of the bridge is not sufficient, the towed load can be pulled across separately.

(3) Track-laying vehicles should be started across so that they do not have to turn.

(4) Brakes should not be used.

Q. When towing over difficult terrain what precautions must be taken? *A.* If possible, apply the brakes on the tow before applying those of the vehicle. Most vehicles designed for towing now provide means to do this.

Q. If a turn is too sharp for a towed load what may be done? *A.* The tow may be uncoupled and negotiated around the bend with winch or block and tackle.

Q. In case a vehicle overturns what is done? *A*. Remove the load and await a maintenance crew with block and tackle and wrecking set.

Q. What is the best way to keep a vehicle from becoming stalled or mired? *A*. Follow a reconnoitered route and make a careful inspection of all doubtful places before attempting to negotiate them.

67. Operation of vehicle not in convoy.—*Q*. When a driver is to make a trip not in convoy, what orders does he receive? *A*. He receives a properly filled out "Driver's Trip Ticket and Performance Record," plus such verbal instructions as may be necessary.

Q. What is the purpose of a "Special Order" directing a driver to complete a certain trip? *A*. In peacetime, the driver of a vehicle not in convoy on an extended trip needs a special order so that he may obtain rations, supplies, fuel, ferriage, etc.

Q. How does the driver obtain rations on a trip not in convoy? *A*. He may take rations in kind, ration with some other organization, or he may be furnished cash in advance in lieu of rations.

Q. How does the driver obtain spare parts or get his vehicle repaired? *A*. If possible, at the nearest Army post; if not, he may have the work done by a local authorized dealer for the type of vehicle he is driving. To provide for the latter case he should be provided before starting with the proper forms showing the method of billing. The regimental transportation sergeant will instruct the driver as to the proper procedure.

Q. How are fuel and lubricants obtained by the individual driver? *A*. From Army posts en route by simply signing issue slips. By the use of tax exemption certificates they may be obtained from private dealers. These certificates must be carried by the driver. In some cases courtesy cards are furnished by oil companies.

Q. In case of accident what does the driver do? *A*.

(1) Renders aid to any injured.

(2) If possible he carefully fills out accident report and obtains names and addresses and statements from all available witnesses.

(3) Notifies local police.

Q. What should the driver do in case the person who ordered the vehicle cannot be found? *A*. He waits a reasonable time at the spot where he was told to report and then reports back to the dispatcher.

Q. Should the individual driver pass a moving convoy? *A*. Not unless ordered to do so by competent authority.

Q. How are locations found in the United States? *A*. By following marked routes with the aid of a road map.

Q. In case the driver feels sleepy what should he do? *A*. Pull to the side of the road and take a rest unless there is an assistant driver to take over.

CHAPTER 10

COMMUNICATION

SECTION I

USE AND CARE OF TELEPHONES

68. Sending, receiving, and recording messages.—*Q.* For the best results, where should the telephone transmitter be placed with respect to the mouth? *A.* Not more than 1 inch from the mouth but not touching it.

Q. How should words be pronounced over the telephone? *A.* Use a moderate tone of voice. Speak slowly and distinctly without slurring any words or syllables. Avoid using words which are difficult to pronounce or whose meaning is not generally known. When necessary to repeat, make the pronunciation more distinct but never shout or raise the pitch of the voice.

Q. How are signals sent? *A.* Singly. Thus 4,370 is sent "four, three, seven, zero." Zero is never pronounced "O." A numeral involving a decimal, like 246.34, is sent thus: "two, four, six, point, three, four." An exact hundred such as 200, is sent: "two hundred"; 4,500, "four five hundred." Even thousands are sent in the same manner, for example, 4,000 is "four thousand."

Q. What is the procedure when the receiver repeats the message back to the sender? *A.* Listen carefully to the message. If any part of the message is incorrectly repeated, call "Error" and repeat that portion of the message. When the message has been correctly repeated back to the sender, the sender should call "Check."

Q. What is the procedure if the sender discovers that he has incorrectly sent part of a message? *A.* He immediately calls "Error" and identifies the portion of the message in error. He then gives the correct message. With short messages it is best for the sender to repeat the entire message.

Q. How are numerals pronounced?   *A.*

Numeral	Pronounced	Principal sounds
0	Ze-ro	Long *o*.
1	Wun	Strong *w* and *n*.
2	Too	Strong *t* and long *oo*.
3	Thuh-ree	Slightly rolling *r* and long *ee*.
4	Fo-wer	Long *o*, strong *w* and final *r*.
5	Fi-yiv	*I* changing from long to short and long *v*.
6	Siks	Strong *s* and *ks*.
7	Sev-ven	Strong *s* and *v*, and well-sounded *ven*.
8	Ate	Long *a* and strong *t*.
9	Ni-yen	Strong *n*, long *i*, and well-sounded *yen*.

Q. What is meant by the "phonetic alphabet"?   *A.*

(1) Certain letters of the alphabet have similar sounds and are often confused in telephone conversations. To avoid this difficulty, the following pronunciation of letters over the telephone is prescribed:

Letter	Spoken as	Letter	Spoken as	Letter	Spoken as
A	Affirm.	J	Jig.	S	Sail.
B	Baker.	K	King.	T	Tare.
C	Cast.	L	Love.	U	Unit.
D	Dog.	M	Mike.	V	Victor.
E	Easy.	N	Negat.	W	William.
F	Fox.	O	Option.	X	X-ray.
G	George.	P	Prep.	Y	Yoke.
H	Hypo.	Q	Queen.	Z	Zed.
I	Inter.	R	Roger.		

(2) The words of the phonetic alphabet are used in place of the letters they represent just as in spelling a word. Expressions such as "A as in Affirm" or "A for Affirm" are not used. For example, in transmitting the words BARTS CHURCH the word BARTS is apt to be misunderstood. The phonetic spelling is as follows: "BARTS, Baker-Affirm-Roger-Tare-Sail." The phonetic alphabet is also used in the transmission by telephone of coded messages. For example, the code group XISV is transmitted as "X-ray-Inter-Sail-Victor."

Q. Give some pointers which will increase the efficiency of receiving messages.   *A.*

(1) Keep the mind on the message; a person cannot receive correctly when he is thinking of something else.

(2) Keep the receiver close against the ear.

(3) Do not interrupt the sender except in cases where not to do so would be of serious disadvantage to the correct reception of the message.

(4) Repeat all messages received. Where messages are long, repeat each sentence as it is sent. When any part of a message is not understood, call "Repeat," and continue to have the message transmitted until it is understood.

Q. How is the telephone answered when it rings? *A.* First: Give the official designation or name of the station. Second: Give the official designation of the person answering.

Q. What is a flash message? *A.* A message used to indicate the approach of aerial targets. The indication of the target is preceded by the word FLASH. repeated three times. and the report is given twice without waiting for an acknowledgment.

Q. Do flash messages follow a particular form? *A.* Yes, they must follow a form and no unnecessary words should be used.

Q. What information is contained in a flash message? *A.*

(Front)

FORM FOR FLASH MESSAGE

(AAAIS)

Organization _____

Serial No. _____   Date _____   How sent_____

Time sent _____   To _____

Observation post	Number of airplanes	Type of airplanes	Time seen or heard	Altitude	Sector in which flying	Direction of flight
1	2	3	4	5	6	7
	One_____	Heavy bombardment.	_____	Very low____	_____	North.
	Two_____	Observation_____	_____	Low_____	_____	NE.
	Three_____	Pursuit_____	_____	Medium____	_____	East.
	_____	Light bombardment.	_____	High_____	_____	SE., south.
OP_____	Several_____	Airplane_____	_____	_____	_____	SW.
	Many_____	_____	_____	_____	_____	West, NW.

NOTE.—Very low—below 500 yards. Low—500 to 2,000 yards. Medium—2,000 to 4,000 yards. High—above 4,000 yards. (Both sender and receiver check off items where possible and save time.)

Q. How are altitudes classified? *A.* High—above 4,000 yards. Medium—2,000 to 4,000 yards. Low—500 to 2,000 yards. Very low—below 500 yards.

Q. What record is made of a flash message? *A.* The sender of the message, the operators who transmit it, and the units which receive it, usually record the message by checking the proper words and filling any appropriate blank spaces on a message form.

69. Laying wire, making connections, and tests.—*Q.* What common types of wire are used in field installations? *A.* Two types, both twisted pair, type W–110 and W–110B.

Q. Describe each type. *A.*

(1) Type W–110 wire has rubber compound insulation covered with weatherproof braid. There are 7 strands, 5 steel and 2 copper. Its weight is 132 pounds to the mile. Resistance is 130 ohms per mile.

(2) Type W–110B is similar to W–110 but has 4 steel and 3 copper strands. Its weight is 132 pounds to the mile. Resistance is 95 ohms

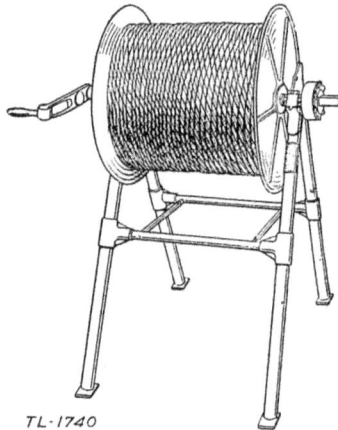

TL-1740

FIGURE 84.—Reel unit RL–31 with reel DR–5.

per mile. (The candidate should be able to identify either type of wire by looking at it. Arrange a pile of short pieces of different types of wire and let the candidate make his own selection.)

Q. What means are provided for laying wire? *A.* The wire is carried or laid by any one of the following means, depending upon the conditions of the roads, terrain, and traffic, and character of hostile fire; motor trucks; especially constructed horse and motor-drawn carts and reels; reel carts, hand-drawn or towed behind communication carts; breast reels; spools or coils carried by hand. If issued on wooden spools, wire may be laid by inserting an iron bar through the spool and paying off from it, or the wire may be rewound onto a spool of special design, provided for the purpose.

Q. What is the present standard reel? *A.* The reel unit RL–31.

Q. Describe the reel unit, RL–31. *A.* It is a portable wire-laying and recovery device. It may be used in any one of several ways, as follows:

(1) Carried litter fashion between two men.

(2) Pushed or dragged along the ground by one man, wheelbarrow fashion, the reel rims acting as wheels.

(3) Mounted inside or on the extended tail gate of any vehicle which provides the required space. Foot fittings are provided for mounting.

(4) Set up on the ground for unreeling or reeling in the wire.

(5) Mounted outside any vehicle by attachment to the outside of the tail gate.

The unit has a removable brake which may be mounted to either end of the axle. Wire is reeled in by means of a crank which may be placed on the end of the axle.

Q. What is the capacity of the reel unit RL–31? *A.* One 1-mile reel (DR–5) or one or two ½-mile reels (DR–4) of wire W–110 or W–110B.

Q. In laying wire should the lines be pulled as tightly as possible or laid loosely? *A.* The lines should be laid loosely in order that the wire may lie flat on the ground, and so as to provide sufficient slack for repairing breaks. At suitable intervals, lines should be attached to objects such as trees or posts in order to leave sufficient slack, and to prevent the wire from being pulled into traffic lanes.

Q. How should a traffic lane be crossed in laying wire? *A.* Where possible the lines should cross roads through the culverts. The wires are passed through the culvert and tied up at the entrance and exit to prevent immersion in the water. When it is necessary to carry the wires overhead they should clear the crown of the road by 14 feet. When a line crosses a road between poles or other vertical supports, the wires should be tied at the base and top of the support on each side of the road. The strain which occurs along the line is met by the tie at the base. If neither of the above methods can be used, the line wires should be buried in a trench, crossing the road at right angles. The wires must be laid snug and well secured at both ends of the trench to prevent their being pulled out.

Q. In laying the wire, at what intervals should it be tested? *A.* It should be tested just after making each splice through the splice when laying wire in the field. In the case of wire on a reel a test should be made before the reel is taken out of storage.

Q. Given two pieces of field wire, describe and illustrate how to make a standard field wire splice. *A.* To obtain a uniform stagger in making the splice, measure back one plier's length (about 6 in.) from the end of one conductor and cut it at this point. Cut one wire of the other pair in the same manner. Crush the insulation on each conductor, starting at about 6 inches from the end and extending back to 2 inches from the end. Use the heel of the pliers for crushing. Score or ring the crushed insulation, at a point about ½ inch from where the crushing began, with the cutting edge of pliers. Using the pliers, skin the crushed insulation off each conductor, being careful not to damage the strands. Clean the strands with the back of the screw driver blade of the electrician's knife. Now tie the long and short conductors together, using a square knot so that the knot occurs about ¼ inch from the insulation. Strip the weatherproof braid from the insulation about ½ inch on each side of the knot. Insert a 6- to 8-inch piece of 19-gage bare copper seizing wire in the knot and pull the knot tight. Bend the seizing wire at the middle and

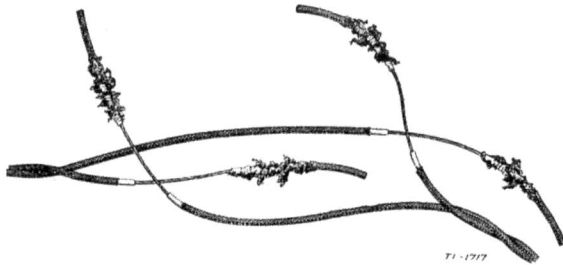

FIGURE 85.—Wires skinned and ready for square knots.

make two or three turns on either side of the knot to bind the ends of the knot. Cut the free ends of the conductors flush with end of insulation. Wrap the seizing wire to left and right of the knot until two turns are taken over the insulation. Cut off the excess wire and press ends of seizing wire into the insulation. Apply two layers of rubber tape followed by two layers of friction tape.

Q. When joints cannot be taped, what should be done to prevent short circuits and grounds? A. The joints should be staggered and raised off the ground.

Q. What telephones are furnished for field use? A. Signal Corps field telephones EE–5, EE–8, and EE–8A.

Q. How are these telephones classified, that is, as local or common battery types? A.

(1) *EE–8.*—Either local or common battery.

(2) *EE–8A.*—Either local or common battery.

(3) *EE–5.*—Local battery only.

TL-1718

FIGURE 86.—Tying square knot.

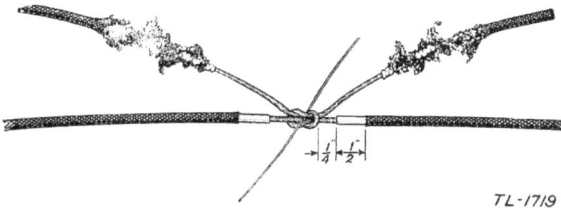

TL-1719

FIGURE 87.—Seizing wire inserted through knot.

Cut tail flush with
rubber insulation

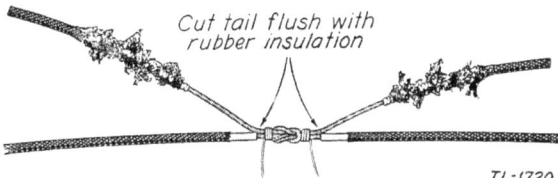

TL-1720

FIGURE 88.—Wrapping seizing wire.

½" rubber   2 turns on rubber

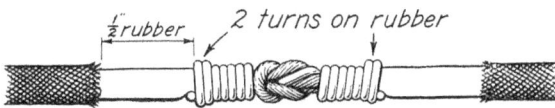

TL-1725

FIGURE 89.—Splice on one conductor after seizing is completed.

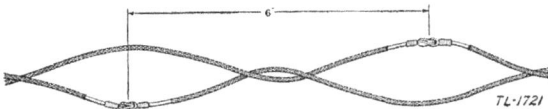

TL-1721

FIGURE 90.—Splice ready for taping.

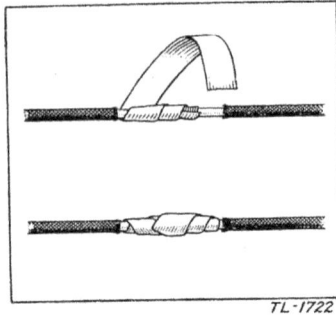

FIGURE 91.—Applying rubber tape.

Q. Name the principal circuits of a local battery telephone. *A.* The primary circuit, which consists of the transmitter, battery, and primary winding of the induction coil. The secondary circuit, which consists of the receiver, condenser, and secondary of the induction coil. The signaling circuit, which consists of the generator and ringer. These circuits are basic in all local battery telephones. Certain telephones, such as the EE–8 and EE–8A, will have additional circuits, but in any case these additional circuits supplement the above basic circuits.

Q. How is the signaling circuit connected? *A.* The circuits of both the generator and the ringer are bridged in parallel across the line terminals.

Q. Does the battery current flow through the signaling circuit? *A.* No. The circuit through the generator is always open except when the crank is turned. A condenser in the ringer circuit prevents the battery current from flowing through the ringer.

Q. How is the station opened when using an EE–8 or EE–8A telephone? *A.* Open the case and remove the handset from the carrying compartment. Place the batteries in the battery compartment.

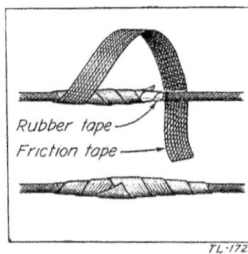

FIGURE 92.—Applying friction tape.

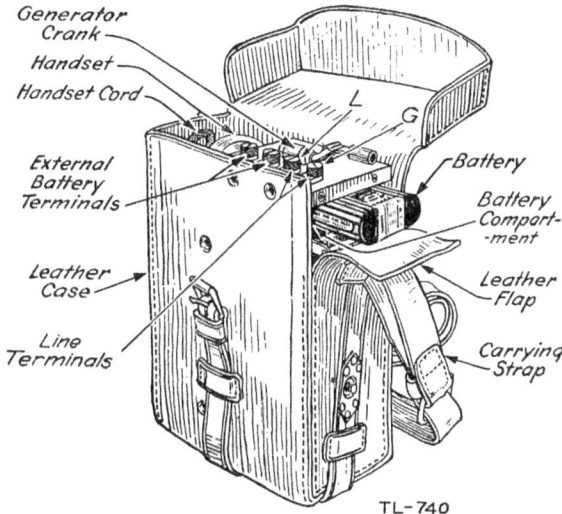

FIGURE 93.—Field telephone EE-5.

Connect the ends of the line to the terminals marked L1 and L2. With a screw driver turn the screw switch to the proper position, depending on whether the telephone is to be operated by local battery or by common battery. There are about 1½ turns of the screw switch between the local battery and common battery positions. If using local battery, call the switchboard, using the generator. If using common battery, removal of the handset from its position on the lever switch will call the switchboard. Report the designation of the station and request a ring back. If using common battery, it will be necessary to replace the handset on the lever switch before the switchboard can ring back.

Q. How is the station opened when using an EE–5 telephone? *A.* Open the cover of the telephone. Remove the crank from the clamp on top and screw it on the magneto shaft which extends out of the side of the case. Remove the handset from the carrying compartment.

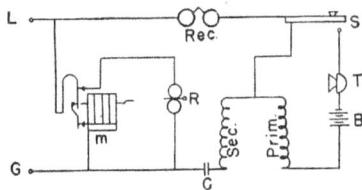

FIGURE 94.—Circuit diagram, EE–5 telephone (modified).

Connect the ends of the line to the terminals **L** and **G**. Call the switchboard by turning the crank. Report the designation of the station and request a ring back.

Q. How is the station closed? *A.*

(1) Report the fact of closing to the switchboard. Disconnect the line from the terminals. With the EE–5 remove the crank and place

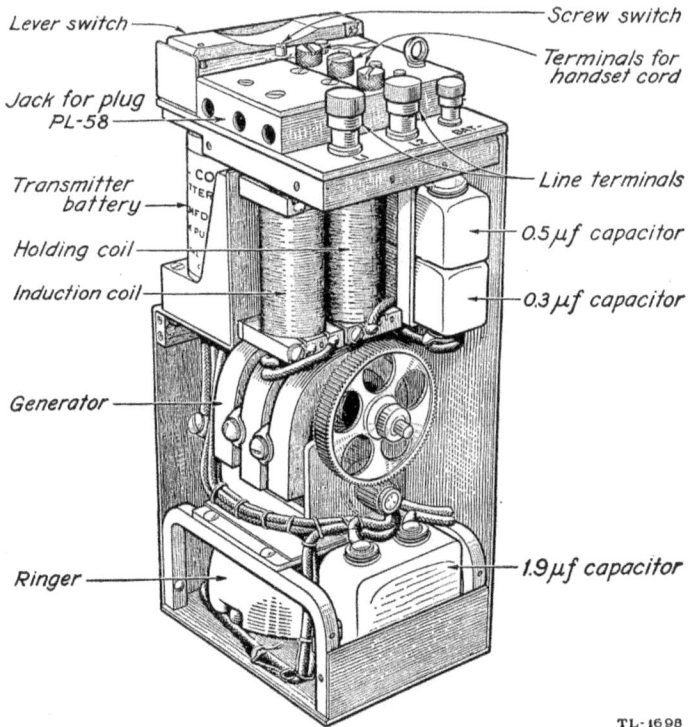

FIGURE 95.—Field telephone EE–8, with side plates removed.

it in its carrying position. If the telephone is not going to be used again *immediately*, remove the battery or batteries.

(2) Wrap the cord about the handset and replace the handset in the carrying compartment.

(3) If the batteries were left in the telephone, be sure that the cord does not operate the handset switch.

Q. Does the current from the generator flow through the receiver and transmitter? *A.* The handset switch must be operated in order to complete the circuit through the transmitter. Hence the generator current will not ordinarily flow through the transmitter. The receiver is always connected across the line, so that generator current could flow through it. However the resistance of the receiver circuit to low frequency currents is very high. The generator produces alternating current at 20 cycles or less per second. Hence very little of the generator current will go through the receiver circuit.

Q. How are the batteries installed in the EE–8 and EE–8A telephones? *A.* Remove the handset from the carrying compartment. Place two batteries BA–30 in the battery compartment (fig. 95), being sure that the bottoms of the batteries rest on the springs and that the tops of the batteries rest against the contacts at the top of the compartment.

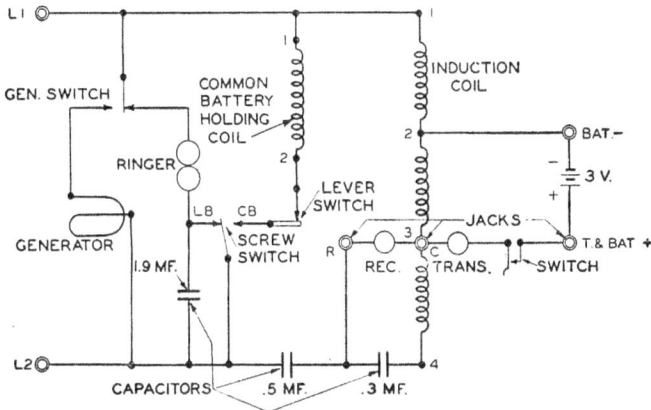

SC-D-1782-D

FIGURE 96.—Circuit diagram, EE–8 telephone.

Q. How is the battery installed in the EE–5 telephone? *A.* One battery BA–9 (4½ volts) is inserted in a spring clip just below the top of the frame (fig. 93). The battery is covered by the leather flap formed by one side of the case. Two screws which hold this flap to the frame must be removed to insert the battery, after which the screws should be replaced.

Q. Demonstrate hooking up two telephones to a length of wire and establish communication (local battery only). *A.* (Practical demonstration proved by actual communication.)

Q. Is it necessary to operate the handset switch in order to listen? *A.* No, and furthermore the operation of the handset switch when listening only is bad practice as it exhausts the battery rapidly.

Q. How may the battery in a local battery telephone be tested?
A. If the battery terminals are touched to the tongue, there should
be a salty taste noted. Another test is to blow lightly into the transmitter while holding the transmitter switch closed. A distinct sound
should be heard in the receiver. In this test, the line should be
disconnected from the telephone.

Q. When should a telephone be tested? *A.* Always before it is
taken out for service. Thereafter the tests are made periodically as
prescribed. The fact that circuits are in constant use is indicative
that they are operating satisfactorily. A telephone which is ordinarily very busy and which suddenly becomes quiet should be tested
at the earliest opportunity. Communication should never be inter-
rupted to make a routine test.

Q. Name the different types of field switchboards. *A.* BD–14,
BD–71, BD–72, BD–9, and BD–11. All types except the BD–14
are monocord switchboards.

Q. Describe the BD–71 and BD–72 switchboards. *A.* The switchboard is inclosed in a plywood case mounted on four collapsible steel
legs. The unit includes switchboard units, cords, operator's telephone with head and chest sets, lights, switches and night alarm,
batteries BA–30, repeating coils, and terminal strips. Outside of the
fact that the BD–71 has only 6-line capacity compared to the 12-line
capacity of the BD–72, there is no practical difference between the
two switchboards.

Q. Describe the BD–9 and BD–11 switchboards. *A.* The BD–9
has a capacity of 4 lines and the BD–11 of 12 lines. The unit consists of a frame on which the individual drops are mounted. Operator's telephone, terminal strips, repeating coils, and similar items
are all separate from the switchboard. A fiber carrying case is provided for the protection of the switchboard when not in use.

Q. Demonstrate hooking up a telephone, a length of wire, and a
BD–71 or BD–72 switchboard. *A.* (Practical demonstration proved
by actual functioning.)

Q. Demonstrate hooking up an operator's telephone, battery, night
alarm, and one line with telephone connected to a BD–9 or BD–11
switchboard. *A.* (Practical demonstration proved by actual functioning.)

Q. What is the purpose of the ground wire? *A.* Protection
against lightning. An air spark gap is incorporated in each unit.
The ground wire grounds one side of the lightning arrester.

Q. Is it desirable for the operator to keep his telephone connected
to two lines which are in use? *A.* No. The extra load which his

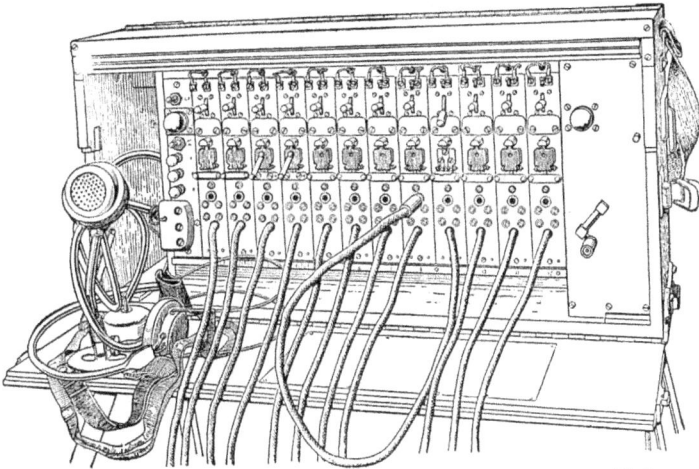

SCL-3

FIGURE 97.—BD-72 switchboard, front view, open.

telephone puts on the lines will impair transmission between the two telephones which are in use.

Q. How does the switchboard operator know when someone is calling the switchboard? *A.* Ringing current on the calling line operates the shutter coil and allows the shutter to drop to the horizontal position. If the night alarm switch is closed, the shutter

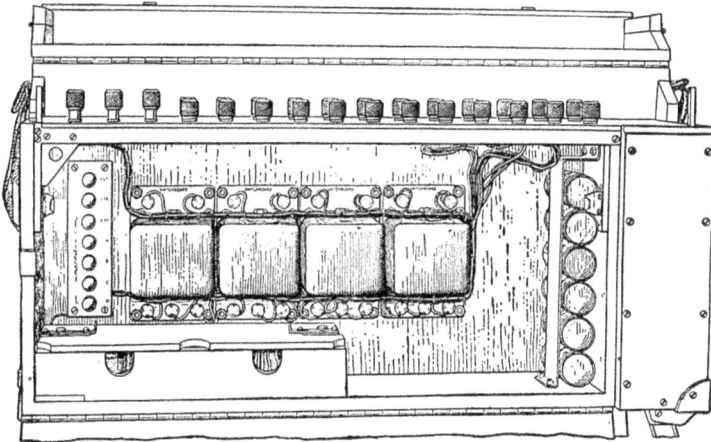

SCL-4

FIGURE 98.—BD-72 switchboard, rear view, open.

COAST ARTILLERY CORPS

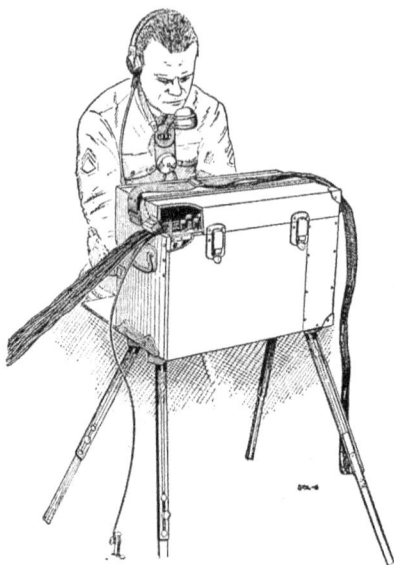

FIGURE 99.—Switchboard BD–71 set up for operation.

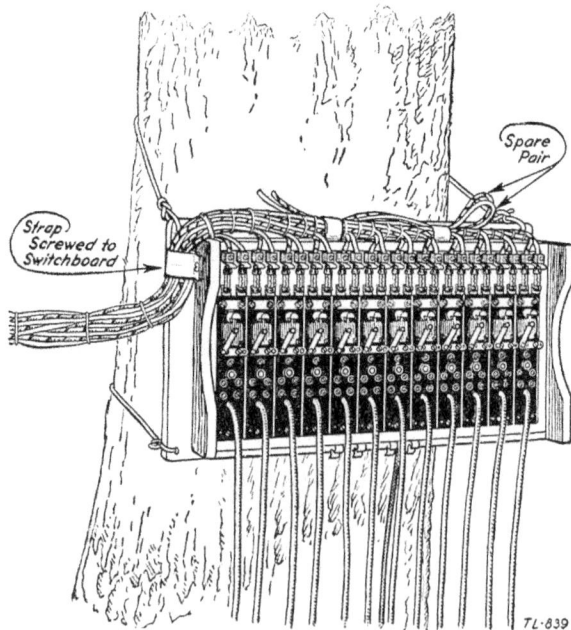

FIGURE 100.—Switchboard BD–11.

cam will close the night alarm circuit and cause the alarm to operate as long as the shutter is down.

Q. What does the operator do when he sees the shutter drop on one of the units? A.

(1) If using a BD–71 or BD–72 switchboard, the operator depresses the ring-talk key on the unit calling. He identifies the switchboard by name and determines the number desired by the party calling. He then restores the calling party's key and rings the called party by raising the ring-talk key on that unit and turning

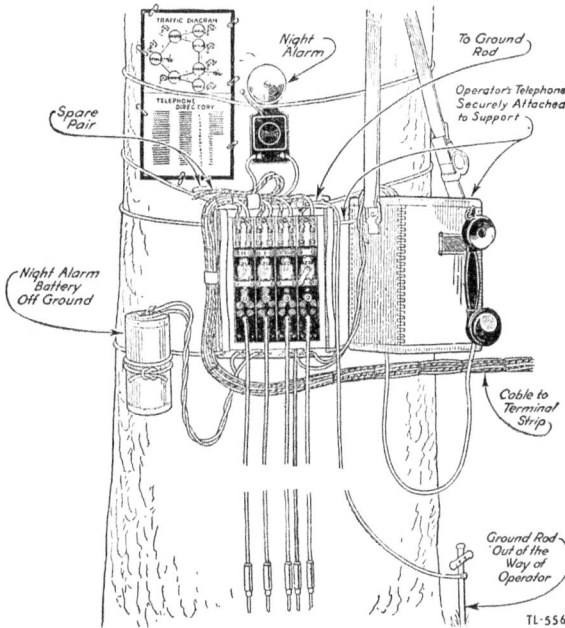

FIGURE 101.—Switchboard BD–9 set up for operation.

the generator handle rapidly several times. He next depresses the called party's ring-talk key to the talk position and inserts the calling party's plug in the jack of the called party. While the ring-talk key is in the talk position the operator's telephone is bridged across the connection, allowing him to supervise the call. The calling party's shutter is left in the dropped position until the call has been completed. A shutter down indicates that the call has not been completed and that further supervision of the connection is necessary.

(2) If using the BD–9 or BD–11, the operator inserts his plug in the calling party's jack, identifies the switchboard by name, and determines the number desired by the calling party. He places the operator's plug in the jack of the called party and turns the generator of the operator's telephone. He then inserts the called party's plug in the jack of the calling party. The operator leaves his plug in the jack until he finishes supervising the call at which time he removes his plug and restores the shutter on the calling party's unit.

Q. Why must the switchboard be upright when in operation? *A.* The shutter drops by gravity. If the board is not upright or inclined slightly forward, the shutter cannot drop.

Q. What are some of the troubles which may occur in a telephone system, and what are the tests and remedies? *A.* See table at end of section.

Q. Describe and demonstrate how to test a telephone. *A.* Install the battery.

(1) Holding the receiver to the ear, blow steadily into the transmitter while alternately operating and releasing the handset switch. The blowing should be very audible as long as the handset switch is at the "on" position. Holding the receiver to the ear, operate the generator. The handle should be easy to turn and the impulses should be heard in the receiver. The ringer should not operate. Short circuit L1 and L2 and turn the generator again. It should now be hard to turn as though a drag had been placed on it, the impulses should be heard in the receiver, and the ringer should not operate. Remove the short.

(2) Connect the telephone to another telephone known to be serviceable. Turn the generator on the other telephone. The ringer of the telephone being tested should operate.

Q. What repairs are telephone operators authorized to make? *A.* With the exception of changing batteries, cleaning contacts which are accessible without taking down the telephone, and changing the headset or handset, the operator is not authorized to make any repairs.

Q. How can most telephone troubles be avoided? *A.* Most of the troubles in telephone communication can be avoided if telephones are carefully used and cared for and are examined and tested before being taken out for service each day. In addition, the batteries must be in good condition, and care must be taken to see that all joints make good contact, including all splices in the lines.

Trouble	Possible cause	Tests and remedies
Home station cannot ring distant station.	(1) Improper line connection at telephone.	(1) Examine connections. Clean and tighten if necessary.
	(2) Open circuit in line.	(2) Examine line for breaks.
	(3) Generator out of order.	(3) Test the telephone at the home station. Repair or replace as may be necessary.
	(4) Receiving circuit open or damaged receiver.	(4) Test the telephone at the distant station. Repair or replace as may be necessary.
	(5) Ringer at the distant station not functioning.	(5) Test the telephone at the distant station. Repair or replace as may be necessary.
	(6) Short circuit in line.	(6) A shorted line is usually distinguished by the generator turning hard. Examine the connections and the line. Remove the short when discovered.
Distant station cannot ring home station.	See above	See above.
Home station can signal distant station but cannot hear distant station talk.	(1) Operator at distant station not operating the handset switch.	(1) Operate the handset switch properly.
	(2) Battery at the distant station dead.	(2) Test the battery. If weak or exhausted, replace.
	(3) Battery contacts corroded.	(3) Examine contacts and battery terminals. Clean if necessary.
	(4) Broken transmitter cord at distant station.	(4) Disconnect handset and touch battery terminals with receiver and transmitter cords, being sure that the handset switch is operated at the same time. A click should be heard in the receiver if the transmitter cord is all right. Replace cord if necessary.
	(5) Handset switch at distant station does not make contact.	(5) Test as in (4) above. Clean and adjust if necessary.
	(6) Carbon in transmitter at distant station packed.	(6) Usually distinguishable by sizzling or crackling noise in receiver. Replace telephone.
	(7) Broken receiver cord at home station.	(7) Disconnect handset. Touch receiver and common cords to terminals of a battery simultaneously. If a click is heard the receiver circuit is all right. Replace cord if necessary.

Trouble	Possible cause	Tests and remedies
Distant station can signal home station but cannot hear home station talk.	See above_____	See above.
Station cannot signal switchboard.	(1) Fuse on switchboard burned out (BD–9 and BD–11 only).	(1) Examine fuses and replace if necessary.
	(2) Shutter stuck on its hinge.	(2) Trip shutter by hand. If the shutter will not drop of its own accord, clean hinge.
	(3) Armature holding shutter is out of adjustment or bent.	(3) (a) *BD–9 or BD–11.*—Hold tip of red (operator's) plug against terminals of section being tested. Ring operator's telephone. If the armature vibrates but does not release the shutter, adjust armature until it does.
		(b) *BD–71 or BD–72.*—Put the plug of an unused circuit across the terminals of the unit being tested. Put ring-talk key to ring position and operate the generator. If the armature vibrates but does not release the shutter, adjust armature until it does.
	(4) Coil of shutter release magnet burned out.	(4) Test as in (3) above. If the armature does not vibrate, the coil is probably burned out. Replace entire unit.
Night bell fails to operate when a shutter drops.	(1) Battery dead_____	(1) Test battery and replace if necessary.
	(2) Loose or dirty connections.	(2) Check through all connections. Clean and tighten where necessary.
	(3) Bell contacts corroded.	(3) Examine bell. Clean the contacts if necessary.
	(4) Shutter dropping does not close bell circuit.	(4) Adjust the contacts so the circuit will be completed.
	(5) Bell coils open___	(5) Connect a receiver in series with a battery and the bell coils. When the circuit is closed there should be a click in the receiver if coils are all right. Replace if necessary.

SECTION II

INSTALLATION AND OPERATION OF REGIMENTAL OR
BATTALION TELEPHONE SYSTEM AND NET

Paragraph

70. Installation and operation of systems and nets.—*Q.* How
are military wire circuits classed according to their use? *A.* As
trunk or local lines.

Q. What is a trunk line? *A.* A line which connects two telephone
switchboards or centrals.

Q. What is a local line? *A.* A line which connects a switchboard
to an individual telephone, or one between two individual telephones.

Q. How are wire circuits classed according to construction? *A.*
As metallic or ground return.

Q. What is the difference between the two? *A.* In the metallic
circuit two wires are used to provide a complete path for the current;
in ground return circuits. the earth replaces one wire.

Q. Are ground return circuits satisfactory? *A.* No, because of
earth currents, cross talk, and the ease with which the enemy can
pick up messages.

Q. What kinds of wire may be used? *A.* Bare wire, insulated
single conductor, insulated twisted pair, and cables may be used.

Q. What kind of wire is generally used for antiaircraft artillery
systems? *A.* Twisted pair, insulated, stranded field wire. Generally W–110 or W–110B wire is used.

Q. What is a line-route map? *A.* A map. map substitute, or
overlay showing the actual routes of wire circuits. It contains also
information as to the location of each headquarters or establishment
served by the system, locations of telephone centrals, switching centrals, test stations. and long locals. Besides showing the actual route
of each wire line, the map shows the type of line construction and
the number of physical circuits in each section of the line.

Q. What is a circuit diagram? *A.* An outline sketch showing
all the circuits and their connections to each other and to the
switchboards. Each circuit is given a number.

Q. In the case of wire lines between higher and lower units, who
is responsible for their installation and maintenance? *A.* The higher
unit is responsible for wire communications from its command post
to the command post of the next lower unit.

Q. What would be a good telephone position? *A.* One free from enemy observation and protected from shell fire. It should not be necessary for the operator to lie down, nor should he be in a position where he is continually annoyed by others.

Q. Where should the switchboard be placed? *A.* At some quiet spot, centrally located, well protected from shell fire.

Q. Give some points to be observed in selecting a route for a telephone line. *A.* It should be as short as possible; avoid main lines of travel; avoid road junctions; follow some natural line of concealment; if in view of the enemy, be camouflaged.

Q. What tests should be made when laying the line? *A.* Before the reels are taken out for laying wire, the wire should be tested for short and open circuits. For this purpose the wire on each reel should be continuous and have the ends exposed. Then as the lines are payed out they should be tested back to the starting point every quarter mile and at every splice made.

Q. What natural features can be utilized in laying a line? *A.* Reverse slopes of hills, trenches, ditches, and woods.

Q. Should there be any slack when wire is laid? Why? *A.* Wire should be laid slack so breaks can be repaired and small pieces cut out if necessary.

Q. How should lines cross a road? *A.* Through a culvert or high enough to be clear of all traffic. Overhead lines must clear traffic arteries or paved roads by 18 feet and other roads by 14 feet.

Q. How are lines laid across water? *A.* Use a single section of weatherproof wire in good condition and weight it sufficiently to hold it against the movement of the current.

Q. What provision is made to maintain communication should a line be broken by shell fire or other causes? *A.* Duplicate lines are run, whenever possible, by different routes, to distant and important OP's or CP's, which are connected to the switchboard and thus readily substituted for defective lines.

Q. How are telephones and operators protected from enemy action? *A.* OP's are normally carefully concealed by camouflage and, if time permits, provided with some cover.

Q. What is "short-stake construction"? *A.* Short stakes about $4\frac{1}{2}$ feet in length are driven into the ground at intervals of about 15 to 20 feet. Small insulated knobs, usually of wood or porcelain, are fastened to the stakes. Wires are attached to these knobs.

Q. What is "wire trench construction"? *A.* Short stakes are driven into the bottom of a small wire trench at intervals of about 15 to 25 feet and wire attached as in the preceding case.

71. Duties of switchboard operators.—*Q*. What are the duties of the switchboard operator? *A*. He should thoroughly understand the board, be familiar with all the connections thereon, be able to make all desired connections promptly, and to make emergency repairs and maintain the board in operation.

Q. What special physical qualifications are essential for a switchboard operator? *A*. He must be able to speak distinctly, must be able to understand speech over the telephone readily, and must be capable of working for prolonged periods under stress.

Q. Can more than two lines be connected for transmitting orders simultaneously to several stations? *A*. Yes. Plug No. 1 may be inserted in jack No. 2, plug No. 2 in jack No. 3, and so on, and the lines so plugged are then connected in parallel.

Q. How are the various stations indicated on the board? *A*. The line number plate (a white celluloid strip) is mounted on each unit. The number or other designation of each station is written in pencil on the corresponding plate. The designation is then easily erased if it is necessary to change it.

Q. Is it necessary to run all the lines of a telephone system through the switchboard? *A*. No, but it simplifies and facilitates control and maintenance of the lines, tracing troubles, substituting spare lines for lines out of order, and keeping in touch with operators at distant observation posts. Also the lines are protected by lightning arresters.

Q. What do telephone operators do when they receive flash messages from observers? *A*. The first operator to receive the message connects up all units which are tied in to his switchboard. Thus a battery operator connects with his battalion operator, who in turn connects with other batteries and with the regiment.

Q. What record is made of a flash message? *A*. The operators receiving it make a record usually by recording the time and checking off the appropriate words and filling necessary blank spaces on a flash message form.

Q. What are some of the phrases prescribed for use by switchboard operators? *A*.

(1) *What number please?*—Used by an operator to request repetition of a number which he had not understood.

(2) *Here's your party.*—Used by an operator whenever it is necessary for him to start the conversation over a connection.

(3) *Waiting?*—Used by an operator in supervising a connection, when no conversation is heard.

(4) *I must interrupt—urgent call from* _____—*please hang up.*— Used by an operator to inform the parties using a circuit that it is required for an urgent call by a certain calling party.

Q. Show how to operate a switchboard.   *A.* Practical demonstration.

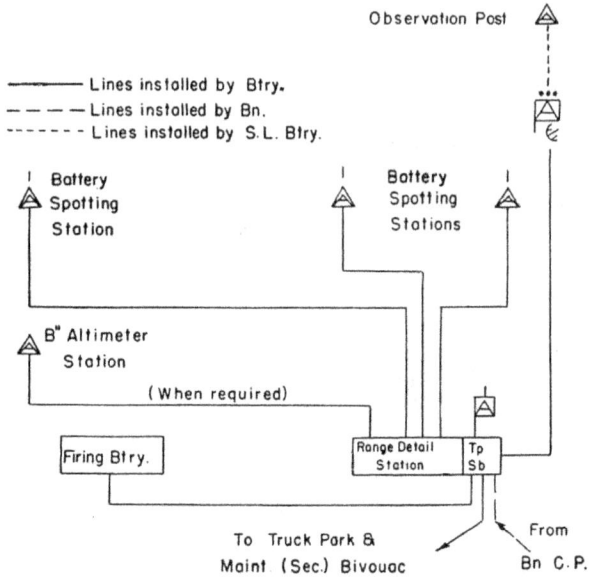

FIGURE 102.—Gun battery telephone net.

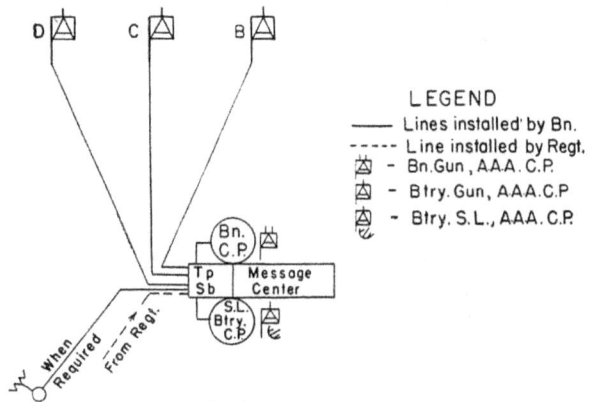

FIGURE 103.—Gun battalion telephone net.

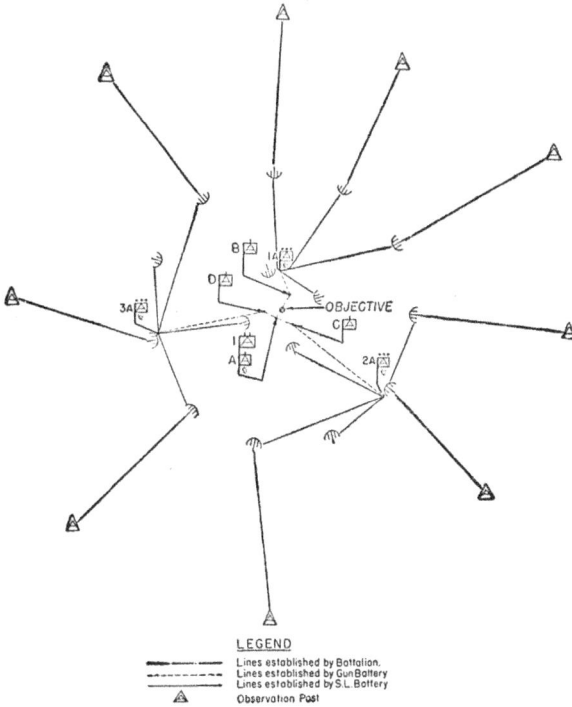

LEGEND
Lines established by Battalion.
Lines established by Gun Battery
Lines established by S.L. Battery
Observation Post

FIGURE 104.—Gun battalion telephone net, including observation posts.

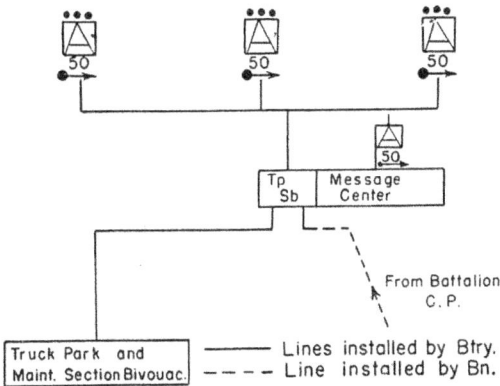

Tp Sb | Message Center

From Battalion C. P.

Truck Park and Maint. Section Bivouac.

——— Lines installed by Btry.
– – – Line installed by Bn.

FIGURE 105.—Machine-gun battery telephone net.

223

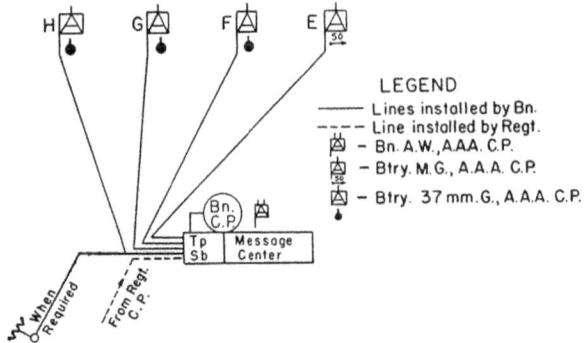

LEGEND
— Lines installed by Bn.
---- Line installed by Regt.
— Bn. A.W., A.A.A. C.P.
— Btry. M.G., A.A.A. C.P.
— Btry. 37 mm.G., A.A.A. C.P.

FIGURE 106.—Automatic weapons battalion telephone net.

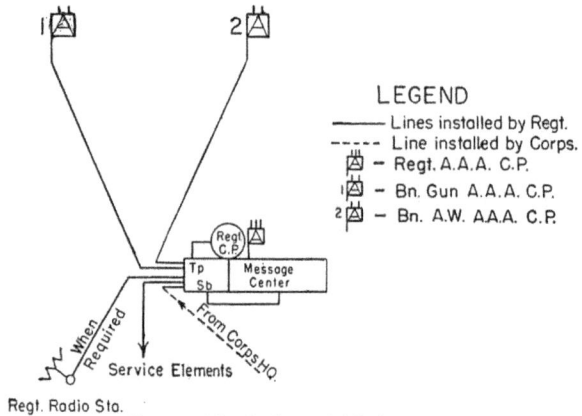

LEGEND
— Lines installed by Regt.
---- Line installed by Corps.
— Regt. A.A.A. C.P.
— Bn. Gun A.A.A. C.P.
— Bn. A.W. A.A.A. C.P.

FIGURE 107.—Regimental telephone net.

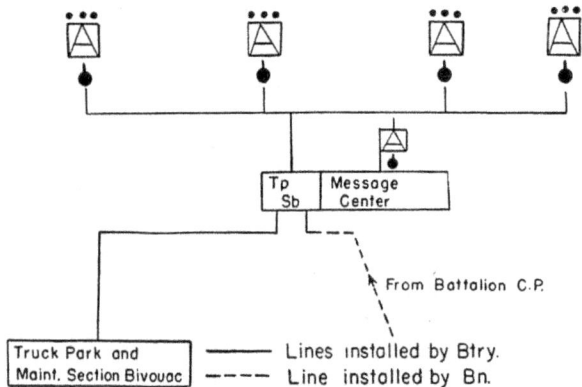

Truck Park and Maint. Section Bivouac

— Lines installed by Btry.
---- Line installed by Bn.

FIGURE 108.—37-mm gun battery telephone net.

Section III

RADIO COMMUNICATION

72. General.—*Q.* What is radiotelegraphy? *A.* Radiotelegraphy is radio communication by means of International Morse Code.

Q. What is radiotelephony? *A.* Radiotelephony is radio communication by means of voice signals.

Q. Radio is used as a means of communication between what units? *A.* All combat units down to and including battalions; individual airplanes, and certain individual vehicles. Under certain situations the battery must also be included.

Q. What are some advantages of radio as a means of communication? *A.* It is independent of roads and traffic and quick to operate. There is no wire to lay or maintain.

Q. What are disadvantages of radio? *A.*

(1) Messages can be intercepted by the enemy.

(2) The number, type, and location of sets in an area may give the enemy an estimate of our dispositions and strength.

(3) A particular frequency or band can be blocked, thereby interfering with communication.

(4) Weather conditions may adversely affect range and quality.

Q. What are important considerations when locating radio stations? *A.*

(1) Radio stations should be in quiet localities protected from weather and enemy fire.

(2) They should not be placed close to sources of possible radio interference such as power lines, telegraph and telephone lines, and other radio stations.

(3) The location should be such that the antenna is in the clear and elevated.

(4) The presence of buildings, hills, woods, and other objects may screen the waves.

Q. What are some of the uses of radio in the military service other than transmission of ordinary messages? *A.*

(1) *Reception.*—(*a*) Location of enemy radio stations on land, sea, or air.

(*b*) Interception of enemy radio traffic.

(*c*) Interception of friendly radio traffic for supervisory purposes.

(*d*) Collection of upper air meterological data.

(2) *Transmission.*—(*a*) Meteorological messages.

(*b*) Time signals.

(*c*) Press reports.

(*d*) Propaganda.

73. Definitions.—*Q.* What is meant by break-in operation? *A.* Break-in operation is operation wherein the receiving operator can interrupt the transmitting operator at any time.

Q. What is meant by call sign? *A.* A call sign is a group of letters, or of letters and numerals, used for station identification.

Q. What is meant by frequency assignment? *A.* The frequency assignment of a station is the frequency or frequencies, usually expressed in kilocycles (kc) or megacycles (mc), at which the station is authorized to operate.

Q. What is meant by heading? *A.* The heading of a message is that part which appears before the text or body begins.

Q. What is meant by intercept station? *A.* An intercept station is a station that copies enemy radio traffic for the purpose of obtaining information, or friendly traffic for the purpose of supervision.

Q. What is meant by internet traffic? *A.* Traffic between stations which are not assigned to the same net is called internet traffic.

Q. What is meant by linking station? *A.* A linking station is a station used for the relay of messages from one net to another.

Q. What is meant by mobile station? *A.* A mobile station is a station that normally operates from a stationary location but which can be rapidly transported to another location.

Q. What is meant by net call sign? *A.* A net call sign is a call sign used to call all stations in a net.

Q. What is meant by position finder station? *A.* A position finder station is a station containing one or more radio receivers capable of finding the location from which incoming radio waves are arriving at the receiver.

Q. What is meant by service? *A.* The service of a message consists of the notations made on a message by transmitting and receiving operators.

Q. What is meant by station log? *A.* Station log is a chronological record of traffic kept at a station.

Q. What is meant by traffic? *A.* Traffic consists of all transmitted and received messages.

Q. What is meant by transmission? *A.* A transmission is a complete communication between stations including all queries, repeat-backs, and receipts.

Q. What is meant by trick or watch? *A.* A trick or watch is a tour of duty as an operator.

Q. What is meant by vehicular station? *A.* A vehicular station is a station so installed in a vehicle that it is capable of operation with the vehicle in motion.

74. Radio nets and net control stations.—*Q.* What is a radio net? *A.* In order that radio communication may follow the proper channels of tactical command, the radio station of a superior unit, and the radio stations of the next subordinate units are grouped together for operation. The superior unit together with its subordinate units, are called a radio net.

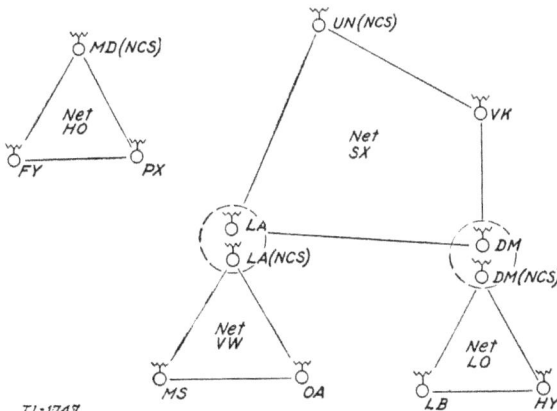

FIGURE 109.—Tactical radio nets.

Q. What are tactical radio nets? *A.* Tactical nets are made up of mobile or vehicular low-powered radio stations of tactical units in the field. They are designated by a name indicative of the superior head-quarters in the net.

Q. How is the transmission of messages usually controlled in a net? *A.* In each net one station is designated as net control station (NCS). The NCS is charged with clearing traffic within the net, working the internet traffic, and maintaining order within the net.

Q. What is a directed net? *A.* In a directed net no station except the NCS can communicate, except for the transmission of urgent messages, with any other station without first obtaining permission of the NCS. A free net on the other hand is not so restricted.

Directed nets are used only when the NCS cannot maintain control otherwise.

Q. How is interference between various nets avoided? *A.* Each net is assigned a definite frequency on which it must operate.

75. Operation.—*Q.* Give some of the transmitting rules that must be observed by an operator. *A.*

(1) An operator will listen on the transmitting frequency assigned his station before making any call or other transmission. If there are other stations working on the frequency, he will not interrupt communication unless such interruption is warranted by the class of his traffic.

(2) All transmissions must be as short and concise as possible. An operator may test his transmitting set before the first transmission by sending a few "V's" followed by his own station call sign.

(3) Messages and transmissions must be sent at a speed which will allow the receiving station to copy them on the first transmission. Thus no transmissions should be faster than the slowest operator in a net can receive them.

(4) Particular care is necessary that all call signs are made slowly and distinctly.

(5) The procedure sign for "Wait" is used when an immediate answer cannot be given.

(6) An "End of message" sign will always be used.

Q. Can the call sign or frequency of another station or net be used? *A.* The use of any call sign or frequency not assigned by higher headquarters is prohibited.

Q. Where will information as to call signs and frequencies be published? *A.* They will be published in Signal Operation Instructions.

Q. What is radio day? *A.* It is the 24-hour period covered by a complete set of station records. It commences at midnight of the time zone in which the station is located, and ends at the following midnight of the same zone. All station records of all stations in the net will be opened and closed in accordance with this rule.

Q. What is the operator's personal sign? *A.* Each operator is identified by a personal sign of one or two letters. No two operators in the same station will use the same sign. It is never transmitted but is used only in keeping of station records.

Q. What are procedure signals? *A.* Arbitrary, nonsecret signals which have a specific meaning. The use of such signals cuts down materially on the time required by operators to handle traffic.

Q. Where are procedure signals prescribed? *A.* In the Joint Army and Navy Radiotelegraph and Radiotelephone Procedure, short title **JANP.**

Q. How may interference by hostile radio stations be minimized? *A.*

(1) By training radio operators in the strict observance of radio discipline and radio security.

(2) By the use of prearranged signals or groups of letters preceding each transmission to identify the station making the transmission.

(3) By frequent changes and limited use of call signs.

(4) By limiting the number of stations in a net.

Q. What messages are transmitted by the radio station? *A.* Only those authorized by competent authority.

Q. If a message is received in code, where is it decoded? *A.* The message center.

Q. What essential elements are recorded in the station log? *A.*

(1) Time entry for each notation.

(2) Operators on duty.

(3) Opening and closing of stations.

(4) Causes of delays in traffic.

(5) Frequency adjustments and changes.

(6) Unusual occurrences such as procedure violations, and verifications.

Q. If the capture of station records by the enemy seems certain, what action should be taken? *A.* They should be destroyed.

Q. State the primary and secondary mission of radio station operators. *A.*

(1) *Primary.*—Closest cooperation with the message center to insure delivery of the message to the addressee without delay and exactly as written by the writer.

(2) *Secondary.*—The keeping of station records.

Q. What benefits are derived from station records? *A.* They are valuable in determining errors made by operating personnel, causes of delays in traffic, and in determining the proper actions necessary for increasing traffic efficiency. They are useful in the recovery of lost messages and as verification records.

Q. What is the range of the set furnished to your organization:

(1) Using radiotelephone?

(2) Using radiotelegraph?

A. See the operating instruction pamphlet issued with the set.

Q. Demonstrate the procedure in setting up a radio set in the field.

Q. Demonstrate the operation of the field power unit.

Q. Demonstrate the operation of the radio set.

Q. Demonstrate ability to send and receive radiotelegraph messages.

CHAPTER 11

SUPPLIES

76. **Supply platoon.**—*Q.* What element of a regiment or separate battalion is responsible for the operation of the supply functions (except ammunition) of the unit? *A.* The supply platoon of the regimental or separate battalion headquarters battery.

Q. Who supervises the operation of the supply platoon? *A.* The unit supply officer.

Q. Of what elements is the supply platoon comprised? *A.* A platoon headquarters and one or more battalion sections. One battalion section is provided for each battalion of the unit of which the supply platoon is a part.

Q. What is the purpose of a battalion section? *A.* It supplies a particular battalion. If a battalion is detached from the .regiment, its battalion section normally accompanies it and functions for the supply of the battalion under the supply officer of the unit to which the battalion is attached.

77. **Mechanics and methods of supply.**—*Q.* How are supplies delivered to the supply platoon when operating in the field? *A.*

(1) They may be delivered to the camp or bivouac of the supply platoon by the train of the next higher administrative unit.

(2) The trucks of the supply platoon may go in convoy to the designated depot or distributing point to secure the supplies.

Q. How are supplies delivered to the batteries? *A.*

(1) The supply platoon trucks may deliver the supplies directly to the batteries.

(2) The supply platoon may establish a distributing point and issue supplies to battery trucks at that point.

(3) Battery trucks may be attached to the supply platoon convoy and go to the depot or distributing point established by the higher administrative unit to receive their supplies.

Q. In the field what written requests are necessary to obtain rations? *A.* In the field no requisition is necessary. A record of the number of men present on the morning report of the regiment is forwarded each day to higher headquarters, and rations are issued daily to supply the number of men reported.

Q. How are gasoline and oil issued in the field? *A.* A reserve of gasoline and oil in containers is carried in each unit. As far as practicable, initial distribution of this reserve is made to each motor vehicle. After the initial issue of gasoline and oil, each motor vehicle operating between army supply points and unit areas replenishes its supply of gasoline and oil at the most convenient class III supply point established by the army. Vehicles operating in forward areas are resupplied with gasoline and oil by exchanging empty containers for full ones brought forward from army supply points either by regimental or divisional transportation.

Q. Name and describe the various field rations. *A.*

(1) *Field ration A.*—This ration corresponds to the peacetime garrison ration and is generally perishable. Being perishable, it is not suitable as a reserve ration.

(2) *Field ration B.*—This ration is the same as field ration A except that nonperishable substitutes replace perishable items. This ration is suitable for reserve purposes.

(3) *Field ration C.*—This ration is a cooked balanced ration in cans. Each ration consists of three cans of prepared meats and vegetables and three cans of crackers, sugar, and soluble coffee. As this ration is not perishable, it is suitable for use as a unit reserve or as an individual reserve.

(4) *Field ration D.*—This ration consists of three 4-ounce chocolate bars per ration. It is a nonperishable ration and is suitable for use as an individual reserve.

Q. For what purpose is ration A employed? *A.* It is issued daily from class I railheads to all divisions and other units not actively engaged with the enemy.

Q. What ration is issued to units engaged in battle? *A.* One of the nonperishable rations or combinations thereof.

Q. What method is sometimes employed by supply officers of regiments (or battalions) engaged in combat to facilitate the issue of rations? *A.* A rear echelon may be established where all kitchens are assembled. From this point trucks are dispatched with cooked meals to locations from which the food can be carried to the troops.

Q. What method is employed to determine the quantities of the various articles of the ration which should be issued by the regiment to the various subordinate units? *A.* The components of the prescribed ration are listed in regulations published either by the War Department or by the commander of the field forces. By using this list of ration components and the strength returns submitted by the various batteries, the regimental or separate battalion supply officer can determine the quantity of each item of the ration to which a unit is entitled.

Q. What practical difficulties may be encountered in attempting to divide the ration articles equitably? *A.*

(1) Meat is received in the form of large cuts such as fore and hind quarters of beef. In dividing up the meat, consideration must be given to the amount of bone and other waste material in different parts of the cut if an equitable division of the meat is to be made. The problem is further complicated by the difficulty of cutting frozen meat.

(2) Sugar, flour, and similar bulk foods will usually be received in 100 pound sacks which must be broken down into smaller quantities for issue to the various units. Care must be taken to reduce losses to a minimum during this handling.

(3) Where the allowances per man are small as in the case of pepper and similar items, or where the quantity of an item to which a unit is entitled is slightly less or greater than the size container in which the item is issued, it may be desirable to issue the item in whole containers, making the necessary adjustment on succeeding days.

(4) In the case of canned articles, equitable division is complicated by the size of the smallest unit of issue, for example, a No. 10 can. This may result in a shortage or overage for a particular unit of a considerable portion of that component. The supply officer may solve this problem by having units provide suitable containers in which portions of the contents of opened cans may be placed, thereby permitting an accurate division of the canned items. If this is not done, units receiving overages will build up a supply of remnants that usually will not be readily usable.

Q. How are supplies secured in peacetime in garrison? *A.*

(1) Rations are drawn by each battery from the commissary.

(2) Clothing is drawn by each battery from the quartermaster warehouse.

(3) Other supplies are drawn from quartermaster, ordnance, engineer, or signal warehouses by the supply officer and issued to the batteries.

78. Handling and storing supplies, including safety precautions.—*Q.* What points should be considered when storing supplies? *A.*

(1) They should be reasonably secure against theft.

(2) They should be protected from the deteriorating effects of weather including excessive heat, excessive cold, or moisture.

(3) They should be conveniently placed and segregated to facilitate handling.

(4) Maximum permissible floor load must not be exceeded.

(5) When supplies must be stored in the open, they should be kept off the ground by placing them on dunnage such as logs, stones, old crates, or any other suitable material available, and kept covered with paulins or other waterproof coverings.

Q. What provision must be made for the storage of inflammable materials? *A.* They must be segregated from other supplies, preferably in a separate building.

Q. How is woolen clothing protected from moths? *A.* When woolen clothing is unpacked and placed on shelves or in bins for issue, sprinkle naphthalene around the clothing in ample quantities. If the clothing is repacked, line the box with wrapping paper and sprinkle naphthalene between the folds of the clothing and between layers. Similar precautions should be taken with woolen blankets.

Q. What safety precautions are necessary when handling and storing gasoline? *A.*

(1) Keep open fires away.

(2) See that gasoline tank trucks have a chain attached which drags on the ground to carry off static electricity which might set fire to the gasoline.

(3) Allow no smoking in the vicinity of places where gasoline is stored or is being handled.

(4) When gasoline is to be stored in cans, the cans of gasoline must be examined for leaks before they are placed in the storehouse.

(5) Special fire-fighting equipment such as sand and carbon tetrachloride types of fire extinguishers should be provided in the storage area.

Q. How should trucks be loaded? *A.* Heavy goods should be on the bottom and toward the rear. Top-heavy swaying loads are dangerous. Sacked goods should be firmly placed and pyramided to prevent shifting and wear from friction. Baled goods should never be placed so as to extend over the sides of the truck. Carefully balanced loads will increase the capacity, as most loads are limited by bulk rather than weight.

Q. Why is the keeping of used rags in storehouses or storage spaces considered to be a dangerous practice? *A.* The rags may become ignited by spontaneous combustion.

79. Capacities of vehicles.—*Q.* Where is the maximum pay load, road and cross country, and the maximum tow load for a particular vehicle shown? *A.* On the vehicle name and caution plate on the dashboard. These loads should never be exceeded except in case of emergency, and then only when specially authorized.

Q. How can overleading of a vehicle be determined when no scales are available and no weights are shown on the cargo? *A.* By noting the position of the rear springs. The driver should be familiar with the appearance of the springs when the truck is carrying its maximum authorized load. Any position of the spring ends below this point indicates that the vehicle is overloaded.

CHAPTER 12

GENERAL SUBJECTS

SECTION I

NOMENCLATURE, ACTION, AND MAINTENANCE OF SMALL ARMS AND THEIR AMMUNITION

80. Rifle, U. S., caliber .30, M1903.—*a. Nomenclature and action.*—*Q.* By what other name is the M1903 rifle often called? *A.* It is popularly referred to as the Springfield rifle, because it was first made at the Springfield Armory, Springfield, Mass.

Q. How would you classify it according to its method of operation? *A.* It is a breech-loading bolt action, magazine rifle.

Q. What is meant by caliber .30? *A.* Caliber .30 means that the distance between two directly opposite lands in the barrel, expressed in inches, is $\frac{30}{100}$ of an inch.

Q. What are the lands and grooves? *A.* The lands are the raised portions of the bore and the grooves are the spaces between the lands.

Q. What direction of twist do the lands and the grooves in the bore give the bullet? *A.* A right twist, or clockwise as seen from the breech.

Q. How does this affect the bullet? *A.* The rotation keeps the bullet from tumbling in its flight, but also causes it to drift slightly to the right.

Q. In firing, should any allowance be made for drift? *A.* No, this is automatically corrected for in the construction of the rear sight leaf.

Q. What ranges can be set on the sight leaf? *A.* Ranges from 100 to 2,850 yards.

Q. What is the weight of the rifle? *A.* About 8¾ pounds.

Q. What is the length of the rifle? *A.* About 43 inches.

Q. What is the muzzle velocity of the ball cartridge? *A.* 2,700 feet per second.

Q. What is the muzzle velocity of the guard cartridge? *A.* 1,200 feet per second.

Q. How many shots can be fired without reloading? *A.* The magazine of the rifle will hold five cartridges and one additional cartridge may be inserted in the chamber, thus making the maximum capacity of the rifle, for any one loading, six shots.

Q. What is meant by the balance of the rifle and where is it located? *A.* As the name implies it is where the rifle balances when held in the hand. It is just below the windage scale and in front of the floor plate.

Q. Point out the following parts:

Barrel.	Slide screw.	Ejector.
Front sight.	Range scale.	Magazine.
Stacking swivel.	Bolt.	Floor plate.
Stock.	Bolt handle.	Guard.
Upper band.	Floor plate.	Trigger.
Lower band swivel.	Sleeve.	Lower band.
Grasping groove.	Firing pin.	Butt swivel.
Hand guard.	Firing pin sleeve.	Butt plate.
Rear sight.	Striker.	Bayonet.
Movable base.	Main spring.	Bayonet guard.
Windage screw.	Extractor.	Bayonet grip.
Windage scale.	Safety lock.	Bayonet catch.
Drift slide.	Cut-off.	Oiler and thong case.
Slide.	Cocking piece.	Brush and thong.

A. See accompanying figures and the rifle itself.

Q. What does the letter "U" on the lower band mean? *A.* If the band is taken off it should be put back with the "U" up, as the band is tapered to fit the barrel and stock.

Q. Explain the working of the extractor. *A.* In loading from the magazine the hook of the extractor catches in the groove on the cartridge case as the follower pushes it up from the magazine. The hook of the extractor continues to hold the cartridge case against the head of the bolt until the bolt is drawn fully to the rear. When the bolt is rotated and drawn to the rear, the extractor brings the cartridge case back with it.

Q. What does the ejector do? *A.* When the bolt is almost fully back, the top locking lug strikes the heel of the ejector and throws the point of the ejector suddenly to the right. As the bolt continues

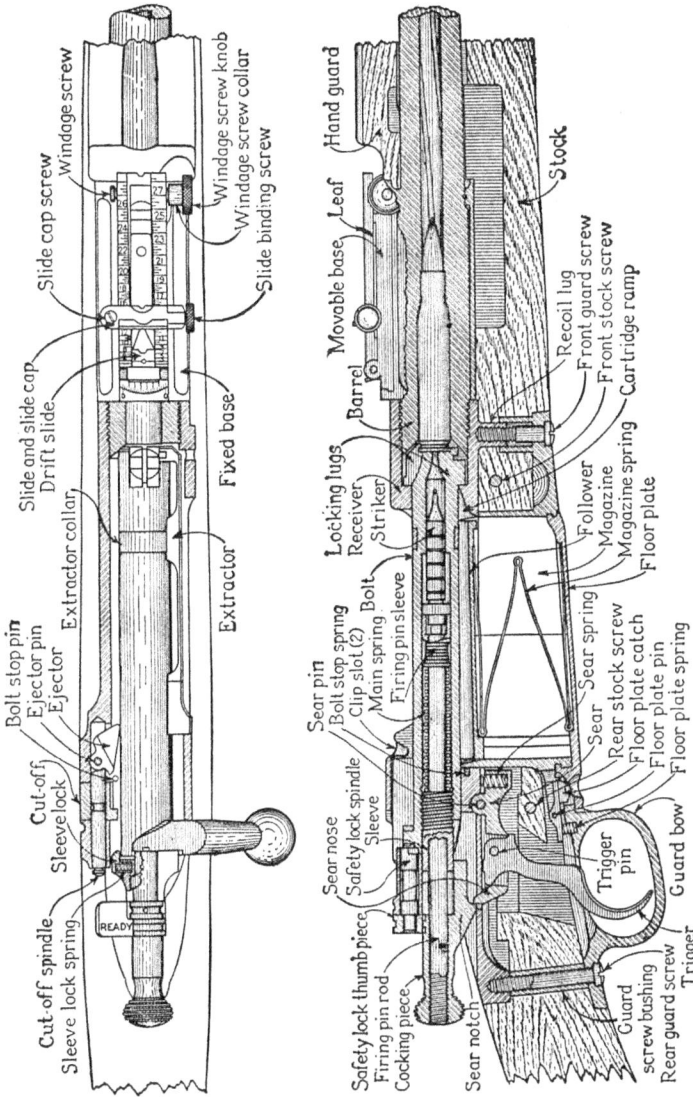

FIGURE 110.—Mechanism of M1903 rifle.

to move back, carrying the cartridge case with it. the ejector hits the rear face of the cartridge case and throws it out of the receiver.

Q. When firing how can you tell when the last cartridge in the magazine has been fired? *A.* After the last cartridge has been fired and the bolt drawn fully to the rear, the follower rises and holds the bolt open to show that the magazine is empty.

Q. Describe the normal open sight. *A.* The top of the front sight appears to be even with the top of the rear sight slide, and the front sight appears in the middle of the rear sight notch.

Q. Describe the normal peep sight. *A.* The top of the front sight appears to be in the center of the peep.

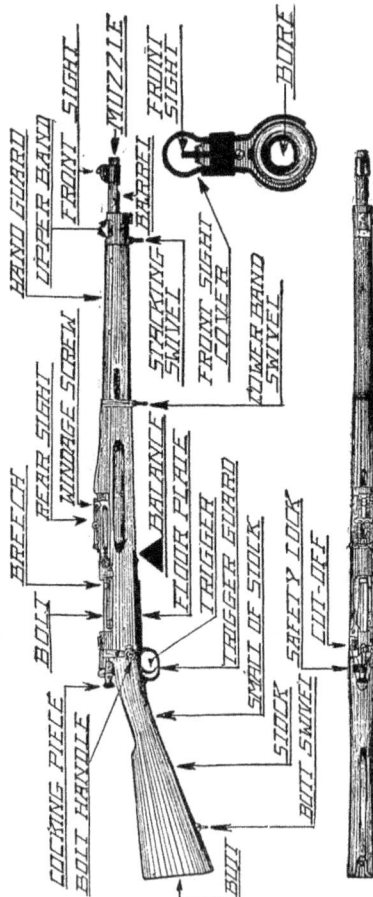

FIGURE 111.—External parts of M1903 rifle.

Q. What is battle sight, and what range is battle sight? *A.* The sight when the sight leaf is *down*—range about 547 yards. The sights are alined as for the normal open sight.

Q. In firing with battle sight, how high is the trajectory above the line of sight at 300 yards? *A.* 2½ feet.

Q. What preparatory instructions should be held before going on the range? *A.* The six steps of preparatory instructions are:

(1) Sighting and aiming exercises (with sighting bar and rest).

(2) Position exercises.

(3) Trigger squeeze exercise.

(4) Rapid fire exercises, in all positions.

FIGURE 112.—Standing position, showing hasty sling adjustment.

(5) Instruction in the effect of wind, sight changes, and use of scorebook.

(6) Examination before going on the range.

Q. What do you mean by the "zero" of a rifle? *A.* The point at which the rear sight must be set for both elevation and windage for any particular range, in order to hit the center of the bull's-eye on a normal day when there is no wind.

Q. What do you mean by "cant" and what is its effect? *A.* It is tilting the rifle to the right or left. The effect is to cause the rifle to shoot low and to the side the rifle is tilted.

239

Q. Where do you focus your eye when aiming a rifle? *A.* On the target.

Q. In firing at a vertical target what is the rule for correcting your fire in elevation? *A.* Square the range expressed in hundreds of yards. The result is the number of inches on the target that the next shot will strike above (or below) if the rear sight is raised or lowered 100 yards. (Example: When firing at the 200-yard range, raising the rear sight 100 yards will move the next shot 4 inches up on the target.)

Q. To shoot to the right (or left), which way would you move the sight? *A.* To shoot to the right move the movable base of the sight to the right. To shoot to the left move the movable base of the sight to the left.

FIGURE 113.—Kneeling position, showing loop sling adjustment.

Q. How much does one point on the windage scale correct for? *A.* Four inches for every 100 yards of range; so at 300 yards range one point corrects for about 12 inches.

Q. How do you figure the effect of a cross wind? *A.* Multiply the range in hundreds of yards by velocity of the wind divided by 10 to find the number of quarter points correction necessary. (Example: When firing at the 200-yard range, a 10-mile wind calls for ½ point correction.)

Q. What is the smallest graduation on the windage scale? *A.* A point—*not* a quarter point.

Q. In what direction do you move the sight to correct for wind effect? *A.* Always into the wind.

Q. How do you aim when using guard ammunition? *A.* Use battle sight and aim at the hips.

Q. What are the positions for rifle firing? *A.* Standing, kneeling, sitting, and prone.

Q. Describe and demonstrate the firing positions. *A.* For all positions face half right from direction of fire and then take the position. The rifle then makes an angle of 45° with the body and should point easily and naturally at the target. The right hand grasps the small of the stock, thumb either around or along the stock; the left hand is near the lower band swivel, piece resting on the palm and in the crotch between thumb and fingers, left elbow as nearly directly under the rifle as possible. The neck and jaw are pressed firmly against the stock. The trigger is squeezed with the first or second joint of the right forefinger. Standing position—feet 12 to 24 inches apart. Kneeling position—the left lower leg is vertical, point of left elbow just over point of knee, the firer sitting on right heel or side of right foot. Sitting position—feet 12 to 24 inches apart and dug into ground, upper arms braced against insides of knees. Prone position—legs straight and well apart, insides of feet flat on ground (or nearly so). shoulders raised on elbows. (See figs. 114 and 115.)

Q. What is the purpose of the sling? *A.* It is used to carry the rifle on long marches, and to afford a steady position and thus improve aim in firing.

Q. How is the sling used in firing? *A.* There are two adjustments called the hasty sling and the loop adjustment (figs. 112 and 113).

Q. What is the most important thing in successful rifle shooting? *A.* Correct trigger squeeze.

Q. Explain how to squeeze the trigger correctly. *A.* The trigger should never be jerked as this always spoils the aim. The rifleman alines his sights accurately on the bull's-eye, and when he has them alined he slowly squeezes the trigger. If the sights wander off the bull's-eye he stops squeezing the trigger, but holds what he has taken up. He brings his sights back into alinement and then continues to squeeze the trigger. The trigger is squeezed only when the sights are on the bull's-eye. After two, or possibly three squeezings the rifle goes off with the sights properly alined. This procedure is the secret of successful rifle shooting.

Q. What mechanisms is the soldier permitted to disassemble? *A.* The bolt and magazine mechanisms only.

Q. Describe how to disassemble and assemble the bolt mechanism. *A.* (1) *To disassemble bolt mechanism.*—(*a*) Place the cut-off at the center notch.

(*b*) Cock the piece and turn the safety lock to a vertical position.

(*c*) Raise the bolt handle and draw out the bolt.

(*d*) Hold bolt in left hand, press sleeve lock in with thumb of right hand to unlock sleeve from bolt, and unscrew sleeve by turning to the left.

(*e*) Hold sleeve between forefinger and thumb of the left hand, draw cocking piece back with middle finger and thumb of right hand, turn safety lock down to the left with the forefinger of the right hand and allow the cocking piece to move forward in sleeve, thus partially relieving the tension of mainspring.

FIGURE 114.—Sitting position.

FIGURE 115.—Prone position.

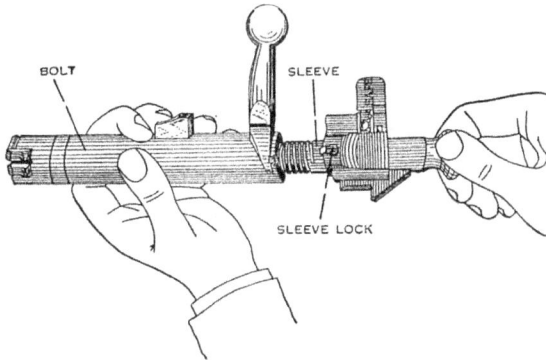

FIGURE 116.—Disassembling bolt mechanism.

(f) With the cocking piece against the breast, draw back the firing pin sleeve with the forefinger and thumb of right hand, and hold it in this position while removing the striker with the left hand.

(g) Remove firing pin sleeve and mainspring.

(h) Pull firing pin out of sleeve.

(i) Turn the extractor to the right, forcing its tongue out of its groove in the front of the bolt, and force the extractor forward and off the bolt.

(2) *To assemble bolt mechanism.*—(a) Grasp with the left hand the rear of the bolt, handle up, and then turn the extractor collar

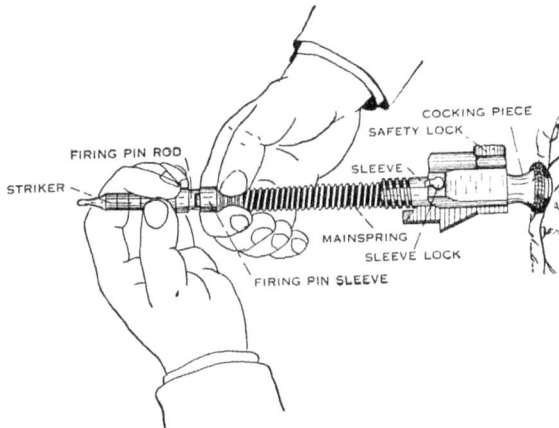

FIGURE 117.—Firing pin.

with the thumb and forefinger of the right hand until its lug is on
a line with the safety lug on the bolt.

(*b*) Take the extractor in the right hand and insert the lug on
the collar in the undercuts in the extractor by pushing the extractor
to the rear until its tongue comes in contact with the rim on the face
of the bolt (a slight pressure with the left thumb on the top of the
rear part of the extractor assists in this operation).

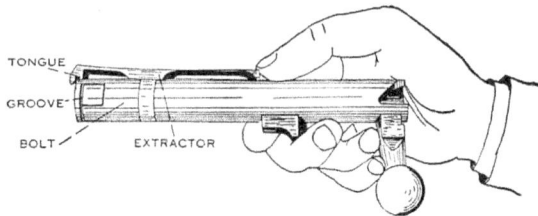

FIGURE 118.—Bolt.

(*c*) Turn the extractor to the right until it is over the right lug.

(*d*) Take the bolt in the right hand and press the hook of the
extractor against the butt plate, or some rigid object, until the tongue
on the extractor enters its groove in the bolt.

(*e*) With the safety lock turned down to the left to permit the
firing pin to enter the sleeve as far as possible, assemble the sleeve
and firing pin.

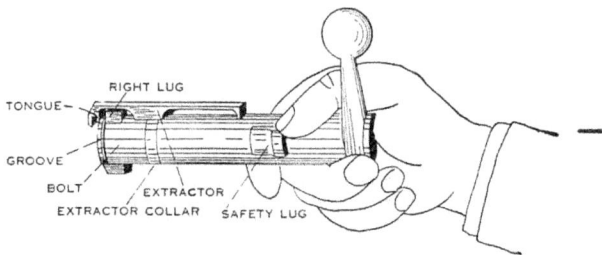

FIGURE 119.—Bolt.

(*f*) Place the cocking piece against the breast and put on mainspring, firing pin sleeve, and striker.

(*g*) Hold the cocking piece between the thumb and the forefinger
of the left hand, and by pressing the striker point against some sub-
stance, not hard enough to injure it, force the cocking piece back
until the safety lock can be turned to the vertical position with the
right hand.

(*h*) Insert the firing pin in the bolt and screw up the sleeve (by turning it to the right) and until the sleeve lock enters its notch on the bolt.

(*i*) See that the cut-off is at the center notch; hold the piece under floor plate in the fingers of the left hand, the thumb extending over the left side of the receiver; take bolt in right hand with safety lock in a vertical position and safety lug up; press rear end of follower down with left thumb and push bolt into the receiver.

(*j*) Lower bolt handle, turn safety lock and cut-off down to the left with right hand.

Q. Describe how to disassemble and assemble the magazine mechanism. *A.*

(1) *To disassemble magazine mechanism.*—With the bullet end of a cartridge press on the floor plate catch (through the hole in the floor plate), at the same time drawing the bullet to the rear; this releases the floor plate. Raise the rear end of the first limb of the magazine spring high enough to clear the lug on the floor plate and draw it out of its mortise; proceed in the same manner to remove the follower.

(2) *To assemble magazine spring and follower to floor plate.*—Reverse operation of disassembling. Insert the follower and magazine spring in the magazine, place the tenon on the front end of the floor plate in its recess in the magazine, then place the lug on the rear end of the floor plate in its slot in the guard, and press the rear end of the floor plate forward and inward at the same time, forcing the floor plate into its seat in the guard.

b. Maintenance of rifle.—*Q.* What causes the most damage to a rifle when it is not properly cared for? *A.* Water and perspiration. If allowed to remain on the metal parts of a rifle rust will form and the surface of the metal will become "pitted."

Q. How is a rifle protected from water and perspiration? *A.* By removing all moisture from the metal parts and covering them with a coating of oil or grease.

Q. Why should a rifle be cleaned after daily drill? *A.* Because handling the weapon removes oil and allows moisture from the hands to get on it.

Q. How do you clean a rifle after daily drill? *A.*

(1) Rub the outside, including the stock and sling, with a rag that has been slightly oiled, and then clean it with a perfectly dry rag. Swab the bore with an oily flannel patch and then with two or three perfectly dry ones. Dust out all screwheads and crevices with a small clean brush.

(2) Immediately after cleaning, swab the bore with a flannel patch saturated with oil (or grease), finally drawing the patch smoothly through the bore and out of the chamber, allowing the cleaning rod to turn with the rifling. Wipe over all metal parts including the bolt mechanism and magazine, with an oily rag and put a few drops of sperm oil on all cams and working surfaces. Put a teaspoonful of linseed oil in the palm of the hand and polish the stock.

Q. Should a rifle be covered when stored in the gun rack? *A*. No. Canvas covers collect moisture which causes the rifle to rust underneath the cover. *The use of rifle covers is prohibited.* (Gun racks will be covered temporarily when barracks are being swept.)

Q. Should a rifle be stored in a gun rack (or any other place) without a protective coating of oil? *A*. No. Even a perfectly clean and dry weapon will soon collect moisture which will damage the metal parts unless they are protected with oil or grease.

Q. How is the sling cleaned? *A*. First wash with a sponge well lathered with castile soap. When partially dry, apply a lather of saddle soap. When this is nearly dry, rub with a dry cloth until the sling is polished. Dry the sling in a cool place. Never dry leather in the sun.

Q. What tool should be used to swab the bore of a rifle? *A*. A barracks cleaning rod should always be used. The thong and brush may be used only if the barracks cleaning rod is not available.

Q. From what end of the rifle should the bore be swabbed? *A*. From the breech, removing the bolt to allow cleaning. Never swab the bore from the muzzle end because of possible damage to the muzzle.

Q. What parts of a rifle should be removed for cleaning? *A*. Front sight cover; floor plate and follower; gun sling; oiler and thong case; and the bolt, which may also be taken apart.

Q. What tool may be used for tightening or loosening screws? *A*. Only a properly fitting screw driver. Never use a bayonet or other substitute because it will damage the screwheads.

Q. Should a rifle be cleaned before firing? *A*. Yes. Always wipe out the bore with a clean patch before going to the firing point. See that no dust, dirt, mud, snow, rags, patches, or other obstructions are in the bore before firing.

Q. What are the three main forms of the residue left in the bore after a rifle is fired? *A*.

(1) A coating of chemicals left by the burned powder.

(2) Particles of unburned or partially burned powder, called *powder fouling.*

(3) Particles of metal from the jacket of the bullet, called *metal fouling.*

Q. How do they damage a rifle if not removed? *A.* The chemicals attract moisture from the air which collects in the bore. The powder fouling and the metal fouling trap moisture underneath, against the bore. The moisture causes rusting and pitting of the bore.

Q. How is a rifle cleaned after firing? *A.* The chemicals and powder fouling are dissolved by scrubbing the bore with a dissolving solution of hot water and issue soap or a sal soda solution. Hot water alone may also be used. (Cold water is used only when none of the other agents are available.)

(1) Remove the bolt and place the muzzle in a vessel containing the dissolving solution. Using a cleaning rod and a flannel patch inserted from the breech, pump the solution back and forth through the bore for about 1 minute.

(2) Next place a brass or bronze wire brush on the rod and run it through the bore all the way down and back three or four times, leaving the muzzle in the dissolving solution. A wire brush is necessary to remove the powder fouling thoroughly.

(3) Next remove the brush from the rod and swab several more times with the dissolving solution.

(4) Then wipe the cleaning rod dry, remove the muzzle from the solution and, using dry clean flannel patches, thoroughly swab the bore until it is perfectly dry and clean. Also dry off the chamber and other metal parts thoroughly.

(5) Finally, inspect the bore for metal fouling. If no metal fouling is present prepare the weapon for storage as you do after daily cleaning and place it in the gun rack. The bore must similarly be cleaned and regreased each day for the next succeeding 3 days to insure that no trace of fouling remains.

Q. How do you inspect the bore for metal fouling? *A.* Hold the butt of the rifle pointed toward the sky and examine the bore from the muzzle, with the eye about 8 inches from the muzzle. If small smears, flakes, or lumps looking like dull lead are seen on the surface of the bore this is metal fouling. It usually occurs within about 6 inches from the muzzle.

Q. What do you do in case you find metal fouling? *A.* Take the rifle to the supply sergeant and ask for instructions.

Q. How is metal fouling removed? *A.* It is removed with metal fouling solution which must be used only by qualified ordnance personnel.

Q. How soon after firing should a rifle be cleaned? *A.* As soon as possible after firing. A weapon should never be put away for the night without being cleaned.

Q. What oils can be used on rifles? *A.*

(1) *For metallic surfaces.*—Sperm oil for lubrication and medium rust-preventive compound for protection from rusting. No other oils should be used unless authorized by the battery commander or his representative.

(2) *For the stock.*—Raw linseed oil. When in the field the stock may be wiped off occasionally with a cloth moistened with sperm oil.

Q. State some of the things one is prohibited from doing with a rifle. *A.*

(1) Except for the removal of those parts permitted for cleaning, a rifle will not be disassembled except by permission of a commissioned officer, and then only under the supervision of a qualified person who knows the provisions contained in the ordnance pamphlet on the subject.

(2) Blued or browned parts of rifles must not be polished.

(3) All mutilations such as carving are prohibited.

(4) Nothing except the authorized oils may be used on a rifle.

(5) Weapons must be unloaded before being taken into barracks or tents.

c. Ammunition.—*Q.* What are the parts of a ball cartridge? *A.* Cartridge case, primer, powder, bullet.

Q. What is the purpose of the primer? *A.* To ignite the smokeless powder.

Q. Describe the bullet for ball cartridge. *A.* It has a core of lead and tin composition inclosed in a jacket of gilding metal, covered with a tin wash. The point is very sharp so as to offer little resistance to the air.

Q. Describe the dummy cartridge. *A.* The bullet is similar to the bullet for the ball cartridge. To distinguish it from the ball cartridge, the dummy cartridge has a tinned case provided with six long straight grooves along it and three holes through it.

Q. Describe the guard cartridge. *A.* The guard cartridge is distinguished from the ball cartridge by having either five grooves around the case (old style) or six short grooves at the shoulder (new style).

Q. What other type of ammunition may be used for guard purposes? *A.* Where the supply of guard cartridges has been exhausted, the gallery practice cartridge M1919 may be issued for guard purposes.

Q. What is the weight of the ball cartridge? *A.* About an ounce; 100 rounds weigh about 5½ pounds.

Q. How is ammunition packed? *A.* In wooden chests (metal lined) containing 1,200 rounds, in cloth bandoleers holding 60 rounds, and metal clips of 5 rounds each.

Q. What types of service ammunition are used with the M1903 rifle? *A.*

(1) Ball, M2 and M1906.

(2) Tracer M1.

(3) Armor-piercing M1.

Q. What distinguishes armor-piercing and tracer ammunition from ball ammunition? *A.*

(1) Armor-piercing is painted black for ¼ inch from the point.

(2) Tracer is painted red for ¼ inch from the point.

Q. At what distance is it dangerous to fire at personnel representing an enemy with rifles loaded with blank ammunition? *A.* Never fire at personnel representing an enemy at distances less than 20 yards.

Q. What is the standard type of ball ammunition? *A.* Ball cartridge, caliber .30, M2, is standard.

81. Rifle, U. S., caliber .30, M1.—*a. Nomenclature and action.*—*Q.* Briefly describe the U. S. rifle, caliber .30, M1. *A.* The U. S. rifle, caliber .30, M1, is a gas-operated, clip-fed, self-loading, shoulder weapon. The gas generated in a cartridge fired in the rifle is utilized to compress the operating rod spring and compensating spring, to extract and eject the fired case, and to cock the hammer. The operating rod spring and compensating spring, which are meantime forcing the cartridges up in the clip, completes the cycle by closing and locking the bolt. The bolt as it goes forward strips the top cartridge from the clip and chambers it. The rifle is then ready to fire.

Q. What is meant by caliber .30? *A.* Caliber .30 means that the distance between two directly opposite lands in the barrel, expressed in inches, is $^{30}/_{100}$ of an inch.

Q. What are the lands and grooves? *A.* The lands are the raised portions of the bore and the grooves are the spaces between the lands.

Q. What direction of twist do the lands and grooves in the bore give the bullet? *A.* A right twist, or clockwise as seen from the breech.

Q. How does this affect the bullet? *A.* The rotation keeps the bullet from tumbling in flight, but also causes it to drift slightly to the right.

Q. In firing, should any allowance be made for drift? *A.* No. This is automatically corrected for in the construction of the rear sight.

Q. How many cartridges may be loaded in this rifle at one time?
A. Eight cartridges are loaded in a reversible clip.

Q. What limits the rate of fire? *A.* The rate of fire is limited only by the proficiency of the soldier in marksmanship, and his dexterity in inserting clips into a magazine.

Q. What is the weight of this rifle? *A.* The weight of the rifle is approximately 9 pounds and the bayonet an additional pound, while the weight of a loaded clip of cartridges (8 cartridges M1) is slightly in excess of 0.5 pound.

Q. What is the maximum range graduation on the rear sight? *A.* The maximum range graduation on the rear sight is 1,200 yards.

Q. What is the muzzle velocity of the ball cartridge? *A.* 2,700 feet per second.

Q. What is the muzzle velocity of the guard cartridge? *A.* 1,200 feet per second.

Q. Point out the following parts:

Butt plate.	Toe.
Rear sight base.	Rear sight nut.
Rear sight elevating knob screw.	Rear sight windage knob.
Rear sight elevating knob.	**Receiver.**
Clip latch.	Front hand guard ferrule.
Bolt.	Front sight screw.
Rear hand guard band.	Gas cylinder.
Rear hand guard.	Stacking swivel.
Lower band.	Stacking swivel screw.
Front hand guard.	Stock ferrule.
Front sight.	Stock ferrule screw.
Gas cylinder plug.	Stock ferrule swivel.
Gas cylinder plug screw.	Gun sling keeper.
Barrel.	Gun sling hook.
Extractor.	Gun sling long strap.
Operating rod.	Gun sling.
Rear sight cover.	Safety.
Aperture.	Trigger.
Stock.	Trigger guard.
Comb.	Gun sling loop.
Heel.	Gun sling short strap.
	Butt swivel.

A. See figure 120.

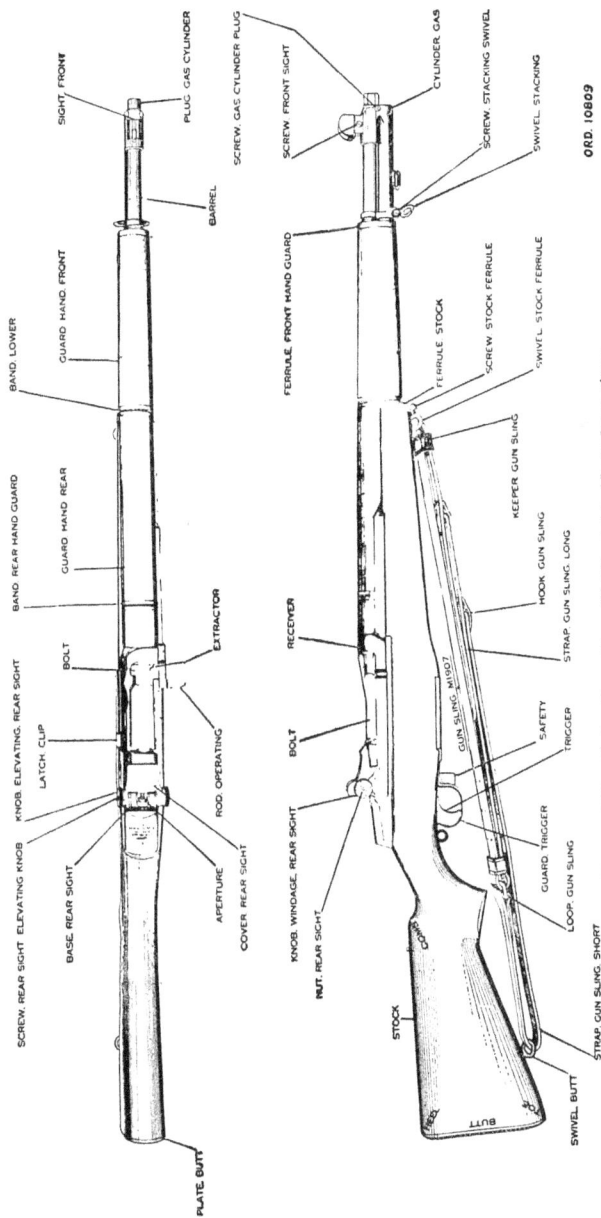

FIGURE 120.—U. S. rifle, caliber .30, M1, right side and top views.

ORD. 10809

Q. How is the ammunition loaded into the cartridge clip? *A.* A clip loading machine (no more to be issued) is sometimes used to load ammunition into clips. In loading the cartridge clip by hand care must be taken to see that the base of each cartridge is close to the rear wall of the clip so that the inner rib of the clip engages the extractor groove in the cartridge, and that each clip is fully loaded with eight cartridges. For ease in inserting the clip it is preferable to have the uppermost cartridge on the right side of the clip.

Q. How is the clip loaded into the rifle? *A.* The operation of loading is performed with the piece locked, that is, with the safety of the piece in its rearmost position, except in sustained firing. Hold the rifle at the balance in the left hand. With the forefinger of the right hand, pull the operating rod handle smartly to the rear until the operating rod is caught by the operating rod catch. With the right hand take a fully-loaded clip and place it on top of the follower. Place the right side of the right hand against the operating rod handle and with the thumb of the right hand press the clip down into the receiver until it engages the clip latch. Swing the thumb to the right so as to clear the bolt in its forward movement and release the operating rod handle. The closing of the bolt may be assisted by a push forward on the operating rod handle with the heel of the right hand.

Q. How is the rifle fired? *A.* The trigger must be squeezed for each shot. After the eighth shot has been fired, the empty clip is automatically ejected upward out of the receiver to the right and the bolt remains open ready for the insertion of another clip. Should the gun be permitted to recoil in such a manner as to cause the trigger to be released and then by a rebound of the shoulder force the gun forward, causing the trigger to strike the trigger finger and be pulled a second time, the firing of a second round may result. *Caution should be exercised against this.*

Q. Can the hammer be cocked without unlocking the bolt? *A.* Yes. In case of misfire or other occasion when it is desired to cock the hammer without unlocking the bolt, unlatch the trigger guard and swing it to its extreme downward position (fig. 121). Close and latch the trigger guard and the rifle is ready to be fired.

Q. How is the rifle set at safe? *A.* The rifle being loaded, if it is not desired to fire at once, is set at safe by pulling the rear on the front surface of the safety until it occupies its rearmost position inside the trigger guard. In this position the trigger cannot be pulled. The rifle may be loaded and operated by hand when the safety is on but it cannot be fired. It can only be set at safe when the hammer is cocked. To set the rifle at ready, push safety to its extreme forward position.

Q. How is a loaded clip removed from the magazine without firing the rifle? *A.* To remove a loaded cartridge clip from the magazine of the rifle without firing, hook the right forefinger over the operating rod handle, pull and hold in the extreme rear position, with the left hand over the magazine, using the left thumb to release the clip latch. The clip with contained cartridges will then be ejected upward out of the magazine into the hand. Do *not* allow the bolt to move forward after pulling it to the rear, as the top cartridge will be moved forward out of its position in the clip, and will prevent the normal ejection.

Q. How is the rifle unloaded? *A.* Pull operating rod handle to rearmost position, thus extracting and ejecting cartridge from the

FIGURE 121.—Cocking hammer without unlocking bolt.

chamber. Hold the operating rod full to the rear and proceed as when removing a loaded clip from the magazine. If it is desired to close the bolt on an empty chamber and retain a partially loaded clip in the magazine, press down on the top cartridge in the clip, allowing the bolt to slide forward, being sure that it is fully closed. *This procedure is exceptional as the rifle is normally either loaded or clear.*

Q. Describe the adjustment of the rear sight. *A.* The rear sight is adjusted vertically by turning the elevating knob, which is on the left side and has numbered graduations for 200, 400, 600, 800, 1,000, and 1,200 yards range. Index lines between the numbered lines correspond to 100, 300, 500, 700, 900, and 1,100 yards range. Adjustment for windage is made by windage knob on the right, each windage graduation representing 4 minutes of angle. The elevating and windage knobs are provided with "clicks" which represent approximately 1 minute of angle or 1 inch at 100 yards. Arrows on knobs indicate direction of rotation for desired changes in point of impact. The rotation of the elevating knob may be eased by forcing the knob outward, away from receiver, while turning.

Q. What safety precautions must be taken with the M1 rifle? *A.* While any cartridges remain in the magazine after a round has been fired, the rifle is ready to fire, and the gun is safe only when it is "cleared." In other words, the gun is never *known* to be safe when the bolt is closed. To clear the gun, pull operating rod fully to the rear, extracting and ejecting cartridge from the chamber, and remove the clip from the magazine *leaving the bolt open.* When the rifle is hot from repeated firing, *a cartridge must not be left in the chamber.* When for any reason firing is suspended for any considerable time *clear the gun.* Overheated cartridges produce abnormal pressures, are liable to preignition, and increase extraction effort to such an extent that the rim of the cartridge case is likely to be pulled off leaving the case in the chamber.

Q. What preparatory instructions should be held before going on the range? *A.* The six steps of preparatory instructions are:

(1) Sighting and aiming exercises (with sighting bar and rest).

(2) Position exercises.

(3) Trigger squeeze exercise.

(4) Rapid fire exercises, in all positions.

(5) Instruction in the effect of wind, sight changes, and use of scorebook.

(6) Examination before going on the range.

Q. What do you mean by the "zero" of a rifle? *A.* The point at which the rear sight must be set for both elevation and windage for a particular range, in order to hit the center of the bull's-eye on a normal day when there is no wind.

Q. What do you mean by "cant" and what is its effect? *A.* It is tilting the rifle to the right or left. The effect is to cause the rifle to shoot low and to the side the rifle is tilted.

Q. Where do you focus your eye when aiming a rifle? *A.* On the target.

Q. What are the positions for rifle firing? *A.* Standing, kneeling, sitting, and prone.

Q. Describe and demonstrate the firing positions. *A.* For all positions, face half right from direction of fire and then take the position. The rifle then makes an angle of 45° with the body and should point easily and naturally at the target. The right hand grasps the small of the stock, thumb either around or along the stock; the left hand is near the lower band swivel, piece resting on the palm and in the crotch between thumb and fingers, left elbow as nearly directly under the rifle as possible. The neck and jaw are pressed firmly against the stock. The trigger is squeezed with the first or second joint of the right forefinger. Standing position—feet 12 to 24 inches apart. Kneeling position—the left lower leg is vertical, point of left elbow just over point of knee, the firer sitting on right heel or side of right foot. Sitting position—feet 12 to 24 inches apart and dug into the ground, upper arms against insides of knees. Prone position—legs straight and well apart, insides of feet flat on ground (or nearly so), shoulders raised on elbows. (See figs. 114 and 115.)

Q. What is the purpose of the sling? *A.* It is used to carry the rifle on long marches, and to afford a steady position and thus improve aim in firing.

Q. How is the sling used in firing? *A.* There are two adjustments, the hasty sling and the loop adjustment (figs. 112 and 113).

Q. What is the most important thing in successful rifle shooting? *A.* Correct trigger squeeze.

Q. Explain how to squeeze the trigger correctly. *A.* The trigger should never be jerked as this always spoils the aim. The rifleman alines his sights accurately on the bull's-eye, and when he has them alined he slowly squeezes the trigger. If the sights wander off the bull's-eye he stops squeezing the trigger, but holds what he has taken up. He brings his sights back into alinement and then continues to squeeze the trigger. The trigger is squeezed only when the sights are on the bull's-eye. After two, or possibly three squeezings the rifle goes off with the sights properly alined. This procedure is the secret of successful rifle shooting.

b. Maintenance of rifle.—*Q.* What causes the most damage to a rifle when it is not properly cared for? *A.* Water and perspiration. If allowed to remain on the metal parts of a rifle rust will form and the surface of the metal will become "pitted."

Q. How is a rifle protected from water and perspiration. *A.* By removing all moisture from the metal parts and covering them with a coating of oil or grease.

Q. Should a rifle be covered when stored in a gun rack? *A.* No. Canvas covers collect moisture which causes the rifle to rust underneath the cover. *The use of rifle covers is prohibited.* (Gun racks will be covered temporarily when barracks are being swept.)

Q. Should a rifle be stored in a gun rack (or any other place) without a protective oil coating? *A.* No. Even a perfectly clean and dry weapon will soon collect moisture which will damage the metal parts unless they are protected with oil or grease.

Q. How is the sling cleaned? *A.* First wash with a sponge well lathered with castile soap. When partially dry, apply a lather of saddle soap. When this is nearly dry, rub with a dry cloth until the sling is polished. Dry the sling in a cool place. Never dry leather in the sun.

Q. Why should a rifle be cleaned after daily drill? *A.* Because handling the weapon removes oil and allows moisture from the hands to get on it.

Q. What tool should be used to swab the bore of a rifle? *A.* A barracks cleaning rod should always be used. The thong and brush may be used only if the barracks cleaning rod is not available.

Q. What tool may be used for tightening or loosening screws? *A.* Only a properly fitting screw driver. Never use a bayonet or other substitute because it will damage the screwheads.

Q. Describe the care and cleaning of the M1 rifle in garrison and camp. *A.* Rifles should be disassembled only to the extent necessary to insure proper condition and cleanliness. The bore of the rifle is always cleaned with a cleaning rod, from the muzzle. If the length of the cleaning rod is such that contact can be made with the face of the retracted bolt, *the bolt must be protected.* To clean the bore push a lightly oiled patch through it and out the breech end. This should be followed with dry patches until several successive ones come out absolutely clean. Push through the bore a patch saturated with oil, to protect its surface. If local climatic conditions necessitate, bores and chambers may be coated with standard rust-preventive compound. To clean screwheads and crevices use a small cleaning brush or small stick. To clean metal surfaces wipe with a dry cloth to remove moisture, perspiration, and dirt. Then wipe with a lightly oiled cloth using aircraft instrument and machine-gun lubricating oil. This protective film must be maintained *at all times.* To clean the outer surface of the rifle, including the stock, hand guards, and sling, wipe off dirt with a lightly oiled cloth, and clean with soft dry one. In cleaning the bore, care must be taken not to foul the cleaning patch in the gas port. Repeat until several successive patches come out absolutely clean. Saturate a patch in sperm oil and push it through the bore, holding the rifle, top up, so that some sperm oil will flow into the gas port.

Q. Describe the care and cleaning of the M1 rifle preparatory to firing. *A.* This differs from the procedure described in the care and cleaning of the M1 rifle in garrison and camp in that Dixon's graphite cup grease No. 3 is substituted for aircraft machine-gun lubricating oil on many of the moving parts of the weapon. The rifle is disassembled and the bore is cleaned and oiled very lightly (*the chamber is not oiled*). Any carbon which may have formed on the gas cylinder plug and the piston head is removed. After thoroughly cleaning and lightly oiling all metal parts a thin uniform coating of the graphite cup grease, referred to above, is applied to the following parts: bolt lugs including locking and operating, bolt guides, cocking cam on bolt, compensating spring, contact surfaces of barrel and operating rod, cam, operating rod guide groove in receiver, and the operating rod spring. *The graphite cup grease should under no circumstances be applied to the follower slide or the under surface of the bolt as the introduction of graphite into the chamber may lead to the generation of excessive pressures.* After the rifle has been assembled all outer surfaces should be rubbed lightly with an oiled rag.

Q. What are the three main forms of residue left in the bore after a rifle is fired? *A.*

(1) A coating of chemicals left by the burned powder and primer.

(2) Particles of unburned or partially burned powder, called powder fouling.

(3) Particles of metal from the jacket of the bullet, called metal fouling.

Q. How do they damage a rifle if not removed? *A.* The chemicals attract moisture from the air which collects in the bore. The powder fouling and metal fouling trap moisture underneath, against the bore. The moisture causes rusting and pitting of the bore.

Q. How do you inspect the bore for metal fouling? *A.* Hold the butt of the rifle pointed toward the sky and examine the bore from the muzzle, with the eye about 8 inches from the muzzle. If small smears, flakes, or lumps looking like dull lead are seen on the surface of the bore this is metal fouling. It usually occurs within about 6 inches from the muzzle.

Q. Describe the care and cleaning of the M1 rifle after firing. *A.* The bore of the rifle must be thoroughly cleaned by evening of the day on which it is fired, and similarly cleaned for the next 3 days. Under no circumstances is the use of any metal fouling solution in the rifle permitted. After disassembling the rifle, the barrel and receiver assembly *with the bolt removed* should be supported at an angle of about 45° with the barrel down. *The bore is always cleaned from the muzzle.* A flannel patch saturated with water is pushed through the bore and out the breech end. This is repeated with several patches followed with dry patches until they come out clean and dry. Then one patch saturated with oil is pushed through the bore and two patches inserted in the slot in the chamber cleaning tool, wrapped smoothly about it, and the chamber scrubbed by twisting the tool. If the rifle is not to be fired the following day, proceed as in care and cleaning in garrison and camp. However, if the rifle is to be used the next day, the procedure described in care and cleaning of the M1 rifle preparatory to firing should be followed. If the rifle is not to be fired for a considerable period, or if local conditions cause excessive formation of rust, a rust preventive should be applied to the bore and chamber after cleaning; for storage all metal parts should be protected in the same manner. Heavy oil and grease must be removed from the bore and chamber before firing.

Q. Describe the care and cleaning of the M1 rifle on the range. *A.* The rifle must never be fired with any dirt, mud, or snow in the bore, and the chamber should be kept free and clean from any oil or dirt. A patch plug or other obstruction must never be allowed to remain in the chamber or bore, and neglect of this precaution may result in serious injury.

Q. Describe the care and cleaning of the rifle in the field. *A.* The rifle must be kept clean and free from dirt, and properly lubricated with graphite cup grease. To obtain maximum efficiency the chamber must be kept clean; additional graphite cup grease is applied to the parts, as prescribed in care and cleaning preparatory to firing, at the first opportunity after indications of excessive friction occur; a light coating of oil is kept on all other metal parts; and carbon is removed from the gas cylinder plug and piston head. In general it should not be necessary to remove any parts of the rifle in the field except the trigger housing group and the gas cylinder plug.

Q. State some of the things one is prohibited from doing with a rifle. *A.*

(1) Except for the removal of those parts permitted for cleaning, a rifle will not be disassembled except by permission of a commissioned officer, and then only under the supervision of a qualified person who knows the provisions contained in the ordnance pamphlet on the subject.

(2) Blued or browned parts of rifle must not be polished.

(3) All mutilations such as carving are prohibited.

(4) Nothing except the authorized oils may be used on a rifle.

(5) Weapons must be unloaded before being taken into barracks or tents.

TABLE OF STOPPAGES

Malfunction	Cause	Correction by soldier
Clip jumps out on seventh round.	Bent follower rod_____	Replace.
Failure to extract_____	(1) Dirty or rough chamber_	(1) Clean chamber.
	(2) Restricted gas port_____	(2) Clean gas port.
Failure to feed_____	(1) Dirty or rough chamber_	(1) Clean chamber.
	(2) Restricted gas port_____	(2) Clean gas port.
	(3) Dirty rifle or improper lubrication.	(3) Clean rifle and lubricate.
	(4) Bent clip_____ _____	(4) Replace clip.
	(5) Ruptured cartridge case in chamber.	(5) Remove ruptured cartridge case.
Fires automatically___	Sear broken or remains in open position.	Replace trigger assembly or hammer spring housing.
Safety releases when pressure is applied on trigger.	Round heel on safety, or broken safety.	Replace safety.
Pull on trigger does not release hammer.	(1) Deformed hammer or trigger or worn trigger pin.	(1) Replace defective part.
	(2) Trigger strikes trigger housing.	(2) Turn in to ordnance.
Hammer releases but gun does not fire.	(1) Bolt not all way seated__	(1) Clean and lubricate.
	(2) Defective ammunition__	(2) Discard round.
	(3) Broken firing pin_____	(3) Replace.
Rear sight elevation jumps.	Loose rear sight nut_____	Tighten.
Creep in trigger_____	Burs on trigger or hammer lugs.	Turn in to ordnance.

c. *Ammunition.*—*Q*. What are the parts of the ball cartridge? *A*. Cartridge case, primer, powder. bullet.

Q. What is the purpose of the primer? *A*. To ignite the smokeless powder.

Q. Describe the bullet for ball cartridge. *A*. It has a core of lead and tin composition inclosed in a jacket of cupro-nickel. The point is very sharp so as to offer little resistance to the air.

Q. Describe the bullet for the dummy cartridge. *A*. The bullet is similar to the bullet for the ball cartridge. To distinguish it from the ball cartridge, the dummy cartridge has a tinned case provided with six long straight grooves along it and three holes through it.

Q. Describe the guard cartridge. *A*. The guard cartridge is distinguished from the ball cartridge by having five grooves around the case (old style) or six short straight grooves at the shoulder (new style).

Q. What other types of ammunition may be used for guard purposes? *A.* Gallery practice cartridge.

Q. What is the weight of the ball cartridge? *A.* About an ounce; 100 rounds weigh about 5½ pounds.

Q. What types of service ammunition are used with the M1 rifle? *A*

(1) Ball, M2 and M1906.

(2) Tracer M1.

(3) Armor-piercing M1.

Q. What distinguishes armor-piercing and tracer ammunition from ball ammunition? *A.*

(1) Armor piercing is painted black for ¼ inch from the point.

(2) Tracer is painted red for ¼ inch from the point.

Q. At what distance must personnel be from the M1 rifle when firing blank ammunition? *A.* Over 20 yards.

Q. What type of dummy cartridge may be used with the M1 rifle? *A.* The corrugated type of dummy cartridge (caliber .30, M1906) may be used for instructional purposes. *The use of the slotted type of dummy cartridge (range, caliber .30, M1) is prohibited.*

Q. What is the standard type of ball ammunition? *A.* Ball cartridge, caliber .30, M2 is standard.

82. Pistol, automatic, caliber .45, M1911 and M1911A1.—*a. Nomenclature and action.*—*Q.* What are the four main requirements for a military pistol? *A.* Accuracy within short ranges; power sufficient to stop an enemy instantly; rapidity of fire; and dependability.

Q. What models of the automatic pistol are used in the military service? *A.* Caliber .45, M1911, and M1911A1.

Q. What markings are on the pistol? *A.* On the right side, "Model of 1911 (or 1911A1), U. S. Army"; on the left side, "United States Property." All pistols are also marked with a serial number.

Q. What is meant by caliber .45? *A.* Caliber .45 means that the distance between two directly opposite lands in the barrel, expressed in inches, is $^{45}/_{100}$ of an inch.

Q. What are the lands and groves? *A.* The lands are the raised portions of the bore and the grooves are the spaces between the lands.

Q. For what use is this pistol intended? *A.* For emergency use at short range.

Q. What is its effective range? Its maximum effective range? Its extreme range? *A.* Its ordinary effective range is 25 yards. Its maximum effective range is placed at 75 yards. Its extreme range, if held at an angle of 30°, is about 1,600 yards.

Q. What is the muzzle velocity of the pistol, and what penetration is obtained? *A.* It has a muzzle velocity of 800 feet per second. A penetration of 1 inch in white pine corresponds to a dangerous wound. At a range of 25 yards this pistol will drive a bullet 6 inches into white pine.

Q. How many shots can be fired without reloading? *A.* Seven.

Q. How fast can it be fired? *A.* Starting with the pistol unloaded it has been fired 21 times in 12 seconds.

Q. Which direction of twist do the lands and grooves in the bore give the bullet? *A.* A left twist, or counterclockwise as seen from the breech.

Q. How does this affect the bullet? *A.* The rotation keeps the bullet from tumbling in its flight, but also causes it to drift slightly to the left.

Q. In firing, should any allowance be made for this drift? *A.* No. At the short ranges at which the pistol is used the drift is so small that it is negligible.

Q. Name the three principal parts of the pistol. *A.* Receiver, barrel, and slide.

Q. Point out the following parts:

Receiver.	Link.
Extractor.	Grip safety.
Barrel.	Link pin.
Ejector.	Safety lock.
Slide.	Barrel bushing.
Firing pin.	Magazine.
Slide stop.	Recoil spring.
Hammer.	Magazine spring.
Rear sight.	Recoil spring guide.
Disconnector.	Magazine catch.
Front sight.	Plug.
Trigger.	

A. See figure 122.

Q. Why is this pistol called the automatic pistol? *A.* Because on being fired, the work of opening the breech, cocking the hammer, extracting and ejecting the empty shell, and forcing a new cartridge into the chamber, is done automatically by the force of recoil.

Q. Describe briefly the functioning of the pistol. *A.* The force of the explosion causes the barrel to recoil slightly. It moves rearward and down until it is stopped by the lug holding link. The slide, having

been unlocked from the barrel, moves to the rear extracting and ejecting the old shell compressing the recoil spring, and cocking the hammer. The greatly compressed recoil spring then forces the slide to the
forward position. During this movement the slide pushes the new

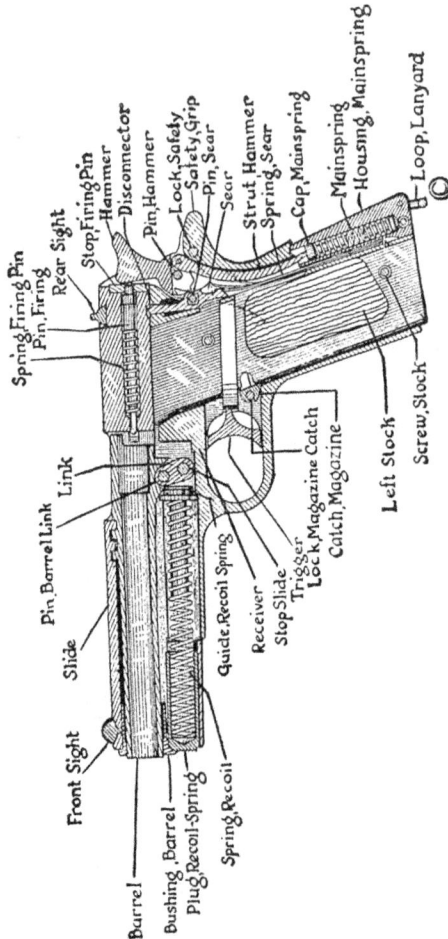

Automatic Pistol, Model 1911.

FIGURE 122.—Automatic pistol M1911.

bullet, which has been raised into the chamber by the action of the
magazine follower into the barrel and pushes the barrel forward
slightly and up, locking into the locking ribs ready for firing again.

Q. How is the pistol loaded? A. A loaded magazine is placed in
the stock and the slide drawn fully back and released, thus bringing

the first cartridge into the chamber. The hammer is thus cocked and the pistol is ready for firing.

Q. How is the pistol fired? *A.* The trigger is pulled, releasing the hammer which falls and strikes the firing pin, driving the latter forward against the percussion primer. This primer ignites the powder which propels the bullet.

Q. How is the pistol loaded again? *A.* The loading is automatic as long as a cartridge remains in the magazine. On recoil the slide is driven to the rear and the recoil spring is compressed. The slide moves forward again, driven by the recoil spring, and another cartridge is carried into the chamber.

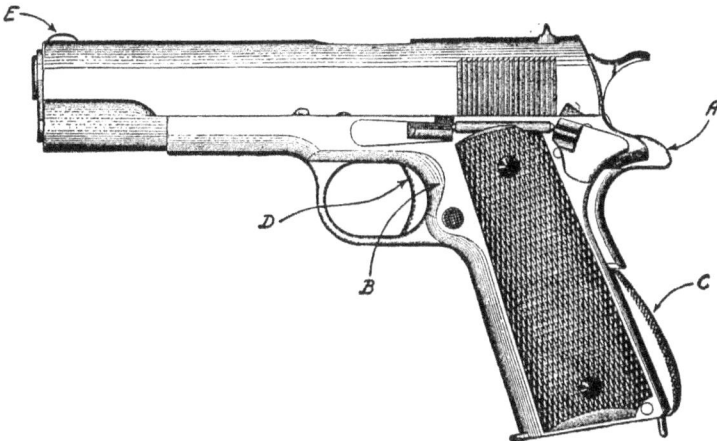

FIGURE 123.—Automatic pistol M1911A1.

NOTE.—Modifications over the M1911 automatic pistol are shown and indicated by letters A, B, C, D, and E.

Q. What are the two automatic safety devices? *A.* The *disconnector*, which positively prevents the release of the hammer unless the slide and the barrel are in the forward position and safely interlocked; this device also controls the firing and prevents more than one shot from following each pull of the trigger. The *grip safety*, which at all times locks the trigger unless the stock is firmly grasped and the grip safety pressed in.

Q. What other safety devices does the pistol have? *A.* The *safety lock*, by which the closed slide and the cocked hammer can be locked positively at will; the *half cock*, which prevents firing until the pistol is fully cocked.

Q. What is the purpose of the locking ribs on the barrel? *A.* To engage in the locking grooves in the slide, thereby securely locking slide to barrel in the firing position, and preventing it from rotating when the pistol is fired.

Q. What is the function of the link? *A.* To pivot the barrel, allowing it sufficient play to rise on its slight forward movement and lock into slide, and also to fall on its rearward movement and disengage from locking grooves, letting the slide continue its movement to the rear. It holds the barrel to its position in the receiver.

Q. What is the function of the magazine follower lip? *A.* When the last bullet in the magazine has been fired, the shell is pushed up into the chamber through the action of the magazine follower spring. As the slide moves to the rear upon recoil the lip engages the pawl on the slide stop and forces it up into slide stop slot on the lower edge of slide, locking the latter in the rearward position. This serves to remind the firer that the last shot has been fired.

Q. What is the difference between the functioning of the extractor and the ejector? *A.* The extractor catches just in front of the shell rim and pulls it back out of the barrel when the slide moves to the rear after firing. The shell strikes against the ejector, which throws it out through the breech opening.

Q. How is the pistol disassembled for cleaning? *A.*

(1) Remove the magazine by pressing the magazine catch.

(2) Press the recoil spring plug inward and turn the barrel bushing to the right until the plug and the end of the recoil spring protrude from their seat, releasing the tension of the spring.

(3) Draw the slide to the rear until the smaller rear recess in its lower left edge stands above the projection on the thumbpiece of the slide stop; press gently against the end of pin of the slide stop which protrudes from the right side of the receiver above the trigger guard, and remove the slide stop.

(4) This releases the barrel link, allowing the barrel, with the link and the slide, to be drawn forward together from the receiver, carrying with them the barrel bushing, recoil spring, plug, and recoil spring guide.

(5) Remove these parts from the slide by withdrawing the recoil spring guide from the rear of the recoil spring and drawing the plug and the recoil spring forward from the slide.

(6) Turn the barrel bushing to the left until it may be drawn forward from the slide.

(7) This releases the barrel which, with the link, may be drawn forward from the slide and by pushing out the link pin, the link is released from the barrel.

Q. How is the pistol assembled? *A.* Proceed in the reverse order. When replacing the slide and barrel on the receiver, care must be taken that the link is tilted forward as far as possible and that the link pin is in place.

Q. How should one grasp the pistol for firing? *A.* To take the grip, hold the pistol in the left hand and force the grip safety device down and back into the crotch formed by the thumb and forefinger of the right hand. The thumb is carried parallel with or slightly higher than the forefinger, it should never be lower. Close the three lower fingers on the stock firmly but not with too tight a grip. The muscles of the arm are held firm but not rigid.

Q. How should one hold the breath while firing? *A.* To hold the breath, draw into the lungs a deep breath, let out a little of the air and stop the rest by closing the throat. Do not hold the breath with the throat open or by muscular effort of the diaphragm.

Q. What should be the position of the body? *A.* The body is a little more than half faced to the left, the feet 12 to 18 inches apart, depending on the man, and the body is perfectly balanced when the pistol is held in the shooting position. The whole position should be natural and comfortable.

Q. How should one squeeze the trigger? *A.* The trigger is squeezed with a steady increase of pressure so as not to know when the hammer will fall. It is squeezed only when the sights are aligned on the target.

Q. How should the sights and the bull's-eye appear when the pistol is fired? *A.* The front sight should appear vertically in the center of the rear sight with its top level with the top of the rear sight. The bull's-eye should appear to rest on the top of the front sight.

b. Maintenance of pistol.—Q. What causes the most damage to a pistol when it is not properly cared for? *A.* Water. If allowed to remain on the metal parts of a pistol it will form rust and the surface of the metal will become "pitted."

Q. How is a pistol protected from water? *A.* By removing all moisture from the metal parts and covering them with a coating of oil or grease.

Q. How is a pistol cleaned after daily drill? *A.*

(1) Rub the outside, including the stock, with a rag that has been slightly oiled, and then clean it with a perfectly dry rag. Swab the bore with an oily flannel patch and then with two or three perfectly dry ones. Dust out all screwheads and crevices with a small clean brush.

(2) Immediately after cleaning, swab the bore with a flannel patch saturated with oil (or grease), finally drawing the patch smoothly through the bore and out of the chamber, allowing the cleaning rod to turn with the rifling. Wipe over all metal parts, including the mechanism and magazine, with an oily rag and put a few drops of sperm oil on all cams and working surfaces.

Q. Should a pistol be stored in a gun rack (or any other place) without a protective coating of oil? *A.* No. Even a perfectly clean and dry weapon will soon collect moisture which will damage the metal parts unless they are protected with oil or grease.

Q. Why should a pistol be cleaned after daily drill? *A.* Because handling the weapon removes oil and allows moisture from the hands to get on it.

Q. What tool may be used for tightening or loosening screws? *A.* Only a properly fitting screw driver. Never use a substitute because it will damage the screwheads.

Q. Should a pistol be cleaned before firing? *A.* Yes. Always wipe out the bore with a clean patch before going to the firing point. See that no dust, dirt, mud, snow, rags, patches, or other obstructions are in the bore before firing.

Q. What are the three main forms of the residue left in the bore after a pistol is fired? *A.*

(1) A coating of chemicals left by the burned powder.

(2) Particles of unburned or partially burned powder, called powder fouling.

(3) Particles of metal from the jacket of the bullet, called metal fouling.

Q. How do they damage a pistol if not removed? *A.* The chemicals attract moisture from the air which collects in the bore. The powder fouling and the metal fouling trap moisture underneath against the bore. The moisture causes rusting and pitting of the bore.

Q. How is a pistol cleaned after firing? *A.*

(1) The chemicals and powder fouling are dissolved by scrubbing the bore with a dissolving solution of hot water and issue soap or a sal soda solution. Hot water alone may also be used. (Cold water is used only when none of the other agents are available.)

(2) Remove barrel of the pistol and place the muzzle in a vessel containing the dissolving solution. Using a cleaning rod and a flannel patch inserted from the breech, pump the solution back and forth through the bore for about 1 minute. Next place a brass or bronze wire brush on the rod and run it through the bore all the way down and back three or four times, leaving the muzzle in the dis-

solving solution. A wire brush is necessary to remove the powder fouling thoroughly. Next remove the brush from the rod and swab several more times with the dissolving solution. Then wipe the cleaning rod dry, remove the muzzle from the solution and, using dry clean flannel patches, thoroughly swab the bore until it is perfectly dry and clean. Also dry off the chamber and other metal parts thoroughly.

(3) Finally, inspect the bore for metal fouling (see below). If no metal fouling is present prepare the weapon for storage as you do after daily cleaning and place it in the gun rack.

Q. How is the bore inspected for metal fouling? *A.* Hold the butt of the pistol pointed toward the sky and examine the bore from the muzzle, with the eye about 8 inches from the muzzle. If small smears, flakes, or lumps looking like dull lead are seen on the surface of the bore this is metal fouling.

Q. What do you do in case you find metal fouling? *A.* Take the pistol to the supply sergeant and ask for instructions.

Q. How is metal fouling removed? *A.* It is removed with metal fouling solution which must be used only by qualified ordnance personnel.

Q. How soon after firing should a pistol be cleaned? *A.* When a pistol has been fired the bore should be cleaned thoroughly not later than the evening of the day on which it is fired. Thereafter it will be cleaned and oiled each day for at least the next three succeeding days.

Q. What oils can be used on pistols? *A.* Sperm oil for lubrication and medium rust-preventive compound for protection from rusting. No other oils should be used unless authorized by the battery commander or his representative.

Q. State some of the things one is prohibited from doing with a pistol. *A.*

(1) Except for the removal of those parts permitted for cleaning, a pistol will not be disassembled except by permission of a commissioned officer and then only under the supervision of a qualified person who knows the provisions contained in the ordnance pamphlet on the subject.

(2) Blued or browned parts of pistols must not be polished.

(3) All mutilations are prohibited.

(4) Nothing except the authorized oils may be used on a pistol.

(5) Weapons must be unloaded before being taken into barracks or tents.

c. Ammunition.—Q. What are the parts of the cartridges? *A.* Cartridge case, primer, powder, bullet.

Q. What is the purpose of the primer? *A.* To ignite smokeless powder.

Q. Of what does the bullet consist? *A.* A core of lead and tin composition inclosed in a jacket of cupro-nickel, weight 230 grains.

Q. What is the weight of the cartridge and bullet complete? *A.* About ¾ ounce; 100 rounds weigh about 4½ pounds.

Q. How is ammunition packed? *A.* The cartridges are packed in pasteboard boxes containing 20 cartridges each. One hundred pasteboard boxes, or 2,000 cartridges are packed in one zinc case, hermetically sealed, with handle for tearing open. The whole is inclosed in a wooden chest, the cover of which is fastened with screw hooks and thumb nuts and sealed.

Q. What types of cartridges are provided for this pistol? *A.*

(1) Ball cartridge, caliber .45, M1911.

(2) Dummy cartridge, caliber .45, M1921.

The dummy cartridge case is tinned and has a ⅛-inch hole in the body.

Section II

CORDAGE AND MECHANICAL MANEUVERS

83. Definitions.—*a. General.*—Rigging involves the technique of handling manila and wire rope and chains in various block and tackle combinations to raise and move heavy loads. It is closely related to the handling of loads by jacks, levers, and similar mechanical devices.

b. Special terms.—(1) *Running end.*—Free end of rope.

(2) *Standing part.*—Whole rope less the running end.

(3) *Paying out.*—Giving slack in rope.

(4) *Bight.*—Loop formed on rope so that the two parts lie alongside each other or across.

(5) *Frapping.*—Drawing together of several turns by passing a rope around all the turns.

(6) *Whipping.*—Wrapping an end tightly with cord or twine to prevent its unlaying when pulled through a pulley or other small opening.

(7) *Unlaying.*—Untwisting of the strands or cords.

(8) *Seizing.*—Lashing the running end back to the standing part.

(9) *Mousing a hook.*—Securing a load held in the hook by wrapping cord or twine across its mouth in such a way as to close it effectively.

(10) *Transom.*—Horizontal spar.

(11) *Upright.*—Vertical spar.

(12) *Belay.*—To make a turn or turns with a running end of rope around a spar, cleat, or the standing part of the rope.

(13) *Thief.*—Knot commonly mistaken for a reef knot, differing in that the end of each rope turns around the standing part, instead of around the other rope.

84. Characteristics of knots.

Name	Use	Directions for tying
1. Overhand	At end of rope to prevent unlaying or to prevent end from slipping through block.	See figure 124.
2. Figure-of-eight	Same as above	See figure 124.
3. Square or reef [1]	To join two ropes of same size.	See figure 124. Pass standing and running parts of each rope through loop of the other in the same direction. Ends of each rope turn around end of other, rather than standing part.
4. Single sheet bend or weaver's. [2]	To join ropes, especially of unequal size.	See figure 124.
5. Double sheet bend [3]	To join ropes of unequal size, especially wet ones.	See figure 124.
6. Two half hitches [4]	To belay or make fast end of rope around own standing part.	See figure 124. End may be lashed down or seized to standing part to prevent slipping.
7. Round turn and two half hitches.	Same as above	See figure 124.

[1] Care must be taken not to tie a thief or granny, as these will slip.
[2] More secure than a reef but more difficult to untie.
[3] More secure than a single sheet bend.
[4] Must not be used for hoisting a spar.

Name	Use	Directions for tying
8. Fisherman's bend or anchor.	To fasten a rope to a ring or anchor.	See figure 124. Take two turns around the iron, then a half hitch around the standing part and between the ring and the turns, then half hitch round standing part.
9. Clove hitch_____	To fasten a rope at right angles to a spar or at beginning of lashing.	See figure 124. If end of spar is free, hitch made by first forming two loops, placing right-hand loop over other, and slipping the double loop over the end of the spar. Otherwise pass end of rope around spar, bring it up to the right of standing part, cross over latter, make another turn around spar, bring up the end between spar last turn and standing part.
10. Timber hitch 5____	To haul or lift spar____	See figure 125.
11. Telegraph hitch___	To hoist or haul spar__	See figure 125.
12. Hawser bend_____	To join two large cables.	See figure 125. Each end is seized to own standing part.
13. Bowline 6_____	To form a loop that will not slip.	See figure 125. Make loop with standing part underneath, pass end from below through loop, over the part, around the standing part, then down through the loop.
14. Bowline-on-a-bight.	To make a comfortable sling for a man.	See figure 125. Make first part as above with double part of rope, then pull bight through sufficiently to allow it to be bent past loop and come up in proper position.
15. Running bowline__	To make a slipknot that will not bind.	See figure 125. Pass end around spar. Form a loop around the standing part with the running end. Make a bowline on

5 Can be easily loosened when strain is taken off, but will not slip under load. When used for hauling spars, a half hitch is added near end of spar.

6 Length of bight depends on purpose for which knot is required.

Name	Use	Directions for tying
		the standing part below the loop—on the running-end side.
16. Cat's-paw_____	To secure a rope to the mouth of a hook.	See figure 126. Form two equal bights; take one in each hand and roll them along the standing part till surrounded by three turns of the standing part; then bring both loops (or bights) together and pass over the hook, and mouse the hook.
17. Sheep shank_____	To shorten a rope or pass a weak spot.	See figure 126. Take a half hitch with the standing parts around the bights.
18. Rolling hitch_____	To haul a larger rope or cable.	See figure 126. Take two turns around the large rope in the direction in which it is to be hauled, and one half hitch on the other side of the hauling part.
19. Blackwall hitch___	To attach a single rope to a hook of a block for hoisting.	See figure 127.
20. Mooring knot_____	To make fast to a mooring or snubbing post.	See figure 127. Take two turns around the mooring or snubbing post, pass the free end under the standing part, take a third turn above the other, pass the free end between the two upper turns.
21. Carrick bend_____	To fasten guys to derricks.	See figure 127.
22. Wall knot and crown on wall.	To finish the end of a rope to prevent unlaying.	See figure 127.

85. Splices.—*Q*. What is the purpose of a short splice? *A*. Short splices are used to join two ropes when an increase in diameter at point of splice is not objectionable.

Q. How is a short splice made? *A*. Unlay the strands of each rope for a convenient length. Bring the rope ends together so that each strand of one rope lies between the two consecutive strands of the other rope. Draw the strands of the first rope along the second and grasp with one hand. Then work a few strands of the second rope over the nearest strand of the first rope and under the second strand, working in a direction opposite to the twist of the rope. Apply the same operation to all strands. Splicing may be continued in the same manner to any extent, and the free ends may be cut off when desired. Splice may be tapered by cutting out a few fibers from each strand each time it is passed through the rope. Splice may be made compact by rolling under a board or under the foot.

Q. What is the purpose of a long splice? *A.* Long splices are used to join two ropes without an increase in diameter at point of splice.

Q. How is a long splice made? *A.* Unlay the rope and bring together as for a short splice. Unlay to a convenient length a strand (*a*, fig. 128) of one rope, laying in its place the nearest strand (*d*) of the other rope. Repeat the operation in the opposite direction with the two other strands (*c*) and (*f*). Lay half of one in place of the unlayed half of the other. Pass the tops through the rope. When the splice has been thoroughly stretched, trim off the ends of the strands.

Q. What is the purpose of an eye splice? *A.* Eye splices are used for fastening a rope to a ring or for making a permanent loop in the end of a rope.

Q. How is an eye splice made? *A.* Unlay a convenient length of rope. Pass one loose strand under one strand of the rope, forming an eye of the proper size. Pass a second strand under the strand of the rope next to the strand which secures the first one. Pass the third strand under the one next to that which secures the second strand. Draw all taut, and continue as for a short splice.

86. Cordage.—*Q.* What is a cord rope? *A.* A rope made of vegetable fibers. These fibers are twisted together to form strands, and several strands are twisted together to form a rope.

Q. What is a wire rope? *A.* A rope of steel, or other metallic wire. A number of wires (usually 19) are twisted into strands and several (usually six) of these strands are laid around a hemp core.

Q. What is the purpose of the hemp core? *A.* To prevent the steel strands from rubbing against and cutting into each other, and to give flexibility to the rope.

Q. How is a rope designated as to size? *A.* By its circumference or diameter in inches.

Q. What is marline, seizing stuff? *A.* Marline and seizing stuff are both small-sized cordage. Marline is usually two stranded and laid up left-handed. Seizing stuff is made of better material and is usually three stranded, right-handed.

Q. How should cord rope be stored? *A.* In coils on skids or blocks so as to permit the circulation of air about the coil. Cord rope should never be stored wet.

Q. How is the strength of cord rope affected when slung over hooks or fastened by knots? *A.* The strength is lowered about one-third.

Q. What care should be used in uncoiling new rope? *A.* Care should be used to find the natural lay of the rope and relieve the twist.

Q. How can rope be identified as right- or left-handed? *A.* By comparing it with a right- or left-handed screw thread.

Q. How should rope be coiled? *A.* Rope should be coiled right- or left-handed according to whether it is right- or left-handed rope.

Q. How should rope be cared for while in storage? *A.* It should be taken out at least once each year, dried, stretched, and all weak spots cut out.

Q. What precaution should be taken before using old rope? *A.* It should be tested, especially when serious damage might result from its breakage.

Q. What should be done before cutting a rope? *A.* A whipping should be placed on each side of the spot where the rope is to be cut. The end of the rope should never be left free to unlay or ravel.

Q. Demonstrate a wall knot, figure-of-eight, bowline, anchor knot, bowline-on-a-bight, sheepshank, cat's-paw, square knot, rolling hitch, clove hitch, blackwall hitch, timber hitch, sheet bend. Explain the use of each. **A.** See paragraph 84.

Q. How is wire rope coiled? *A.* Small size wire rope may be coiled in the same manner as cord rope. Large wire rope should be coiled in a figure eight.

Q. What is the principal precaution to be taken in using a wire rope? *A.* Never let it become kinked while under a strain.

Q. Name the component parts of a clip. *A.* The roddle, the **U**-bolt, and the **U**-bolt nuts.

Q. How should wire rope be attached? *A.* Normally wire rope should be attached with thimble and clips. The rope may be secured around the thimble by splicing, but this requires expert work. In the absence of clips a seizing of wire may be used.

Q. What precautions must be taken when using wire rope attached with thimble and clips? *A.* See that the roddles of all clips are in contact with the *long* end of the rope (fig. 125). After the wire has been subjected to strain the clip bolts must be tightened.

87. Blocks and tackle.—*Q.* What is a block? *A.* A block consists of a shell or frame of metal, or wood and metal, housing a grooved pulley or sheave on which rope runs, and giving support to the ends of a pin on which the sheave revolves. A hook, usually free to revolve (swivel), may be attached to one end of the block and often an eye or becket to the other end.

Q. What are the parts of a block? *A.* The parts of a block are the shell or frame, the sheave or wheel upon which the rope runs, and the pin upon which the wheel turns in the shell.

Q. How are blocks designated? *A.* Blocks are designated by the length of the shell in inches and by the number of sheaves. Those with one, two, three, or four sheaves are called single, double, triple, and quadruple. The smallest size of block (length in inches) that will take a given rope is nine times the rope diameter. Self-lubricating blocks should be used where obtainable.

Q. Define the following:

(1) *Snatch block.*—A single block with the shell open at one side to permit the insertion of a rope without passing the end of the rope through the block.

(2) *Running block.*—A block that is attached to the object to be moved.

(3) *Standing block.*—A block that is fixed to some permanent object.

(4) *Simple tackle.*—Consists of one or more blocks rove with a single rope.

(5) *Return.*—Each part of the rope between the two blocks, or between either end and the block, is called a return.

(7) *Overhaul.*—To separate the blocks.

(8) *Round in.*—To bring the blocks closer together.

(9) *Chock-a-block.*—When the blocks are in contact, the tackle is said to be chock-a-block.

Q. What is the purpose of blocks? *A.* Blocks are used to change the direction of pull and to give mechanical advantage. A man of average weight will pull about 60 pounds horizontally.

Q. What is a tackle? What are the different parts? *A.* A tackle is a rope and block or a combination of ropes and blocks working together for use in lifting or moving objects. The standing part of a tackle is that part of the rope between the end which is made fast to one of the blocks, to the weight to be moved, or to some fixed point and the point where it passes over the first sheave. The running part of the rope consists of all the parts moving between the sheaves. The fall is that part of the rope to which the power is applied. A moving block is called a running block and a fixed block is called a standing block.

Q. What is meant by the power or mechanical advantage of a tackle? *A.* The ratio of the load to the power required to lift or move it. Thus if a man weighing 150 pounds can just lift a 600-pound weight with a certain tackle, the power of the tackle is four.

Q. What mechanical advantage is gained by the use of blocks? *A.* In simple tackles the mechanical advantage gained is a direct function of the number of ropes supporting the load. Thus, if the movable block is a double one, then four ropes will sustain the load and the mechanical advantage gained is four.

Q. Draw sketches showing: a whip tackle; a whip on a whip; a runner; a gun tackle; a luff tackle; and show the power of each. *A.* See figure 136.

Q. Why is a runner a more powerful tackle than a whip? *A.* Because the pull is in the same direction as that in which the load is moved instead of in the opposite direction.

Q. Rig a whip tackle. Gun tackle. Luff tackle.

Q. What is a chain or triplex block? *A.* A chain or triplex block consists of a train of gears operated by a large wheel over which an endless chain passes. Power is applied to this chain. The gears operate a sprocket wheel over which runs a heavy chain, the links of which fit into the sprockets. The heavy chain lifts the weight and is provided with a hook for supporting the weight. Chain blocks are rated according to their lifting capacities and range by ½-ton changes from 1 to 5 tons.

88. Slings.—*Q.* What are slings made of? *A.* Slings are made of manila rope, wire rope, or chains. The most common is a manila sling made by splicing the two ends together.

Q. How is a sling used? *A.* To use a sling, pass it around the article to be lifted. Pass the bight formed by one end through the bight formed by the other and then over the lifting hook. If the sling is the same size as the lifting rope, it should make a minimum angle of 30° with the horizontal. At this angle, the stress in each branch of the sling is equal to the stress in the lifting rope. If the angle is greater than 30°, the load is limited by the strength of the lifting rope; if less than 30°, by the strength of the sling.

Q. How do you make a barrel sling? *A.* To sling a barrel horizontally, make a bowline with a long bight. To sling a barrel vertically, make an overhand knot on top of the two parts of the rope; open out the knot and slip each half of it down the sides of the barrel; secure with a bowline.

89. Lashings.—*Q.* How should two spars be lashed at right angles? *A.* Make a clove hitch around the upright a few inches below the transom. Bring the lashing under the transom, up in front of it, horizontally behind the upright down in front of the transom, and back behind the upright at the level of the bottom of the transom and above the clove hitch. Keep the following turns outside the previous ones on one spar and inside on the other, not riding over the turns already made. Make four more turns. Make two frapping turns between the spars, around the lashing, and finish the lashing off either around one of the spars or any part of the lashing through which the rope can be passed. Do not make the final clove hitch around the spar on the side toward which the stress is to come, as it may jam and be difficult to remove. While tightening, beat the lashing with a handspite or pick handle. This is called a square lashing.

Q. How should two spars be lashed for a pair of shears? *A.* Lay the two spars alongside each other with the points below which the lashing is to be made resting on a skid. Make a clove hitch around one spar, and take the lashing loosely eight or nine turns about the two spars, above the clove hitches, without riding. Make two or more frapping turns between the spars, and finish the lashings off with a clove hitch above the turns on one of the spars. Open the butts of the spars and pass a sling over the fork. Hook or lash a block to this sling. Make fast fore and back guys with clove hitches to each spar just above the fork.

Q. How should three spars be lashed for a gin or tripod? *A.* Mark on each spar the location of the center of the lashing. Lay two of the spars parallel to each other with an interval a little greater than the diameter of a spar. Rest their tips on a skid and lay the third spar between them with its butt in the opposite direction so that the marks on the three spars will be in line. Make a clove hitch on one of the outer spars below the lashing and take eight or nine loose turns around the three. Take a couple of frapping turns between each pair of spars in succession and finish with a clove hitch on the central spar above the lashing. Pass a sling over the lashing and the tripod is ready for raising.

90. Gins and shears.—*Q.* Describe a gin. *A.* A gin is a tripod of poles or spars. The two outside poles are called legs and the third called the pry pole. A gin requires no guys.

Q. What is a gin used for? *A.* For lifting heavy weights vertically.

Q. Name the different parts of a garrison gin. *A*. Two legs, pry pole, bolt and clevis, windlass and ratchet, two handspikes, three shoes, two braces, and tackle.

Q. How much can be safely lifted with it? *A*. About 17,000 pounds.

Q. Explain briefly how a garrison gin is assembled and raised. *A*. The legs and pry pole are laid on the ground with the heads together and in position for assembling. The head is then assembled by putting the pin through the legs, pry pole, and clevis. The windlass is put in place and the braces are brought up and put in their places. The gin is raised, after assembly, by raising the head and bringing up the foot of the pry pole towards the feet of the two legs (see fig. 134).

Q. Describe the shears. *A*. Shears consist of two spars, of a size suitable for the weight to be raised, lashed together at the fork. A tackle is fastened to the lashing by a strap or otherwise, the hook is moused, and holdfasts are required.

Q. What are shears used for? *A*. Shears are used for lifting heavy weights to move them a short distance, as in loading or unloading a ship or railroad car.

Q. How are shears held in position after being raised? *A*. By means of guys (lines from the top of the shears to holdfasts on the ground).

Q. How are the shears raised? *A*. If not too heavy, lift the head and haul in on the proper guys. If too heavy to raise in this way, form a crutch by lashing together two poles near their upper ends, the feet of the crutch being slightly in rear of the heels of the shears and secured to prevent them from slipping. Lay the rear guy over the crutch and raise the crutch by means of two light guy ropes, until it is inclined at an angle of about 45° to the front. Haul on the rear shear guy, allowing the crutch to rise as the shears rise. After the shears are raised high enough so that the crutch ceases to act, it is lowered by means of its guy ropes. Footings should be prepared for heavy shears on hard ground, and the legs should be connected by a lashing to prevent spreading.

Q. How is a load moved horizontally by means of shears? *A*. By slacking off on one guy and taking up on the other. Tackle may be used for this purpose if necessary.

91. Anchorages.—*Q*. What is the purpose of an anchorage? *A*. It furnishes a holdfast for the tackles or guy cables in handling heavy loads by means of tackles, gins, shears, etc.

Q. Describe two forms of anchorages. *A.* The picket holdfast is a succession of pickets driven into the ground in continuation of the guy or cable and at right angles in a vertical plane to the line of pull, connected from the top of one picket to the bottom of the next, with the direct pull on the bottom of the first picket. A deadman is a log, rail, or other arrangement buried in the ground, horizontally at right angles to the line of pull, which is applied to the center of the deadman.

Q. What is the purpose of a holdfast? *A.* Holdfasts are used to anchor a line to the ground, as for a guy.

Q. How is a holdfast made? *A.* Drive stout pickets into the ground, one behind the other, in the line of pull. Secure the head of each picket, except the last, by a lashing to the one behind it. Tighten the lashings by rack sticks, and then drive the points of these into the ground to hold them in position. The distance between pickets should be several times the height of the picket above the ground. A single good ash picket, 3 inches in diameter, driven 5 feet into good solid earth, will stand a pull of about 700 pounds.

Q. What is the purpose of a deadman? *A.* A deadman has the same use as a holdfast except that it has greater strength, but requiring more labor to construct.

Q. How is a deadman prepared? *A.* Lay a log or timber in a transverse trench with an inclined trench intersecting it at its midpoint. Pass the cable down the inclined trench, take several turns around the log, and fasten the cable to the log by half hitches and marline stopping. If the cable is to lead horizontally or incline downward, pass it over a log at the outlet to the inclined trench. If the cable is to lead upward, the log is not necessary, but the deadman must be buried deeper. The strength of the deadman depends upon the strength of the log and holding power of the earth.

Q. How can you determine the holding power of a deadman? *A.* For given cable pull, the number of square feet of deadman bearing surface required is determined by dividing the total pull to be placed on the deadman by the value given for the depth and cable inclination selected. (See table LXVI, FM 5–10.)

92. Jacks.—*Q.* Name two types of jacks. *A.* Screw jacks and hydraulic jacks.

Q. What is the usual maximum lift of a screw jack? *A.* Usually from 16 to 18 inches. Care should be taken that it is not screwed too high.

Q. What liquid is used for filling hydraulic jacks? *A.* A mixture of alcohol and water.

Q. How would you determine what mixture of fluid to put into a hydraulic jack? *A.* Consult the manufacturer's handbook or operational instructions.

Q. Can all hydraulic jacks be used in both the horizontal and vertical positions? *A.* No. They are manufactured in two classes, horizontal jacks and base jacks. Horizontal jacks may be used in any position. Base jacks are used in the upright position, but may be inclined provided that the head is always kept higher than the base.

Q. How may the two classes of jacks be distinguished? *A.* They may be distinguished by the fact that the base jack has the pump and reservoir within the ram while the horizontal jack, which is the shorter one, has a separate piece for the cylinder which has no connection with the reservoir except through the pump and the lowering passage.

Q. How is the hydraulic jack filled? *A.* To fill the hydraulic jack start with the ram down. Remove the lowering valve and hexagonal cap. Fill through the large hole. Small amounts necessary to replace liquid which has leaked out may be put in by removing the small screw and filling.

Q. How is the hydraulic jack emptied? *A.* To empty, have the ram down, place the finger over the escape hole in the cylinder, pump the ram until the bottom of it is above the hole, then open the lowering valve and remove the finger from the escape hole allowing air to enter under the ram. The ram may now be lifted out. Remove the lowering valve and hexagonal cap and invert the jack to allow the liquid to run out.

Q. What general precautions should be taken in using the hydraulic jack? *A.* The ram should be kept down when not in actual use. In raising a weight the lever should be inserted in the socket with the projection down. The lowering valve should be closed. The lever should be worked up and down with a slow steady stroke. A weight is lowered by opening the lowering valve. The speed of lowering is controlled by the valve. It should be lowered slowly and never checked suddenly. The jack is designed to lift its rated load with one man operating the lever.

Q. What general precautions should be taken in using the screw jack? *A.* It must never be screwed out to the full extent in raising a weight. The threads must be kept clean, lubricated, and free from burs. The jack should not be used to lift weights greater than its rated capacity.

93. Blocks and wayplanks.—*Q.* What are the requirements of blocks? *A.* Blocks should be sound, free from knots, unpainted, and free from grease. Edges should not be splintered or rounded.

Q. What precautions should be used in erecting a crib? *A.*
(1) The foundation should be level.
(2) Large enough blocks should be selected.
(3) The blocks should then be laid crossing each other in alternate tiers, and the weights supported should be made to bear equally upon all sides of the base.

Q. What is a wayplank and how is it used? *A.* A wayplank is a hard plank, preferably of oak, usually about 15 feet long, 12 inches wide, and 3 inches thick. Each end is beveled for a distance of 6 inches, the bevel on one end being on the side opposite the bevel on the other end. These planks are used chiefly for forming temporary tramways for rollers, or for the wheels of carriages bearing heavy weights, especially in crossing weak bridges.

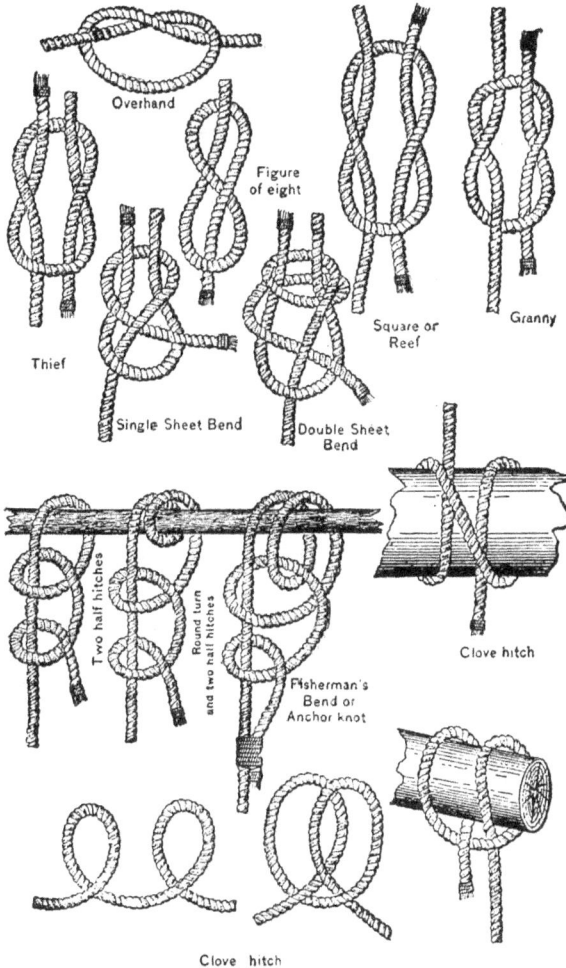

FIGURE 124.—Types of knots—hitches, bends, overhand, etc.

281

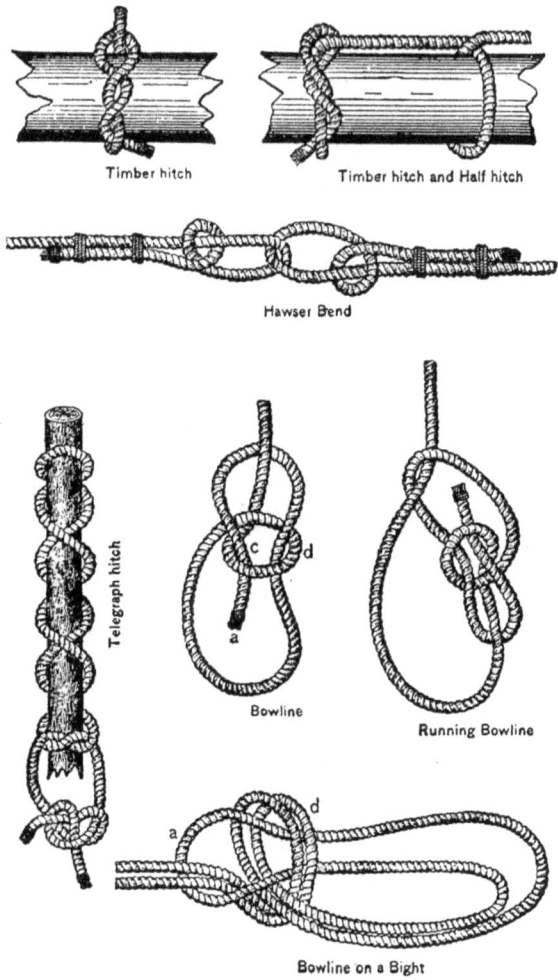

FIGURE 125.—Types of knots—hitches, bends, and bowlines.

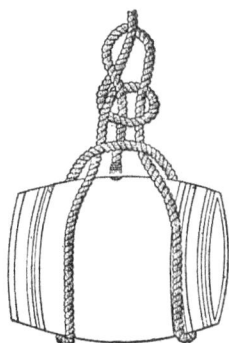

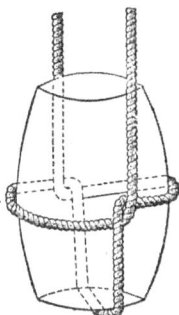

Sling for barrel horizontal.   Sling for barrel vertical.

Cat's Paw a

Rolling Hitch   Sheepshank   Cat's Paw b

FIGURE 126.—Miscellaneous knots.

Blackwall Hitch

Mooring Knot

Carrick Bend

Wall Knot

Wall Knot

Crown on Wall

FIGURE 127.—Miscellaneous knots and hitches.

Short Splice.

Short Splice.

Short Splice.

Long Splice.

Long Splice.

Eye Splice.

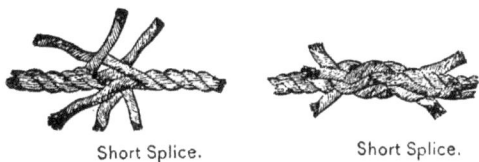

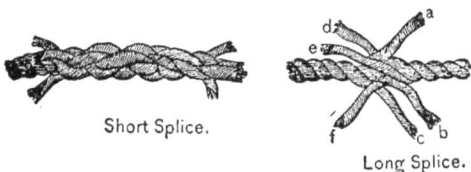

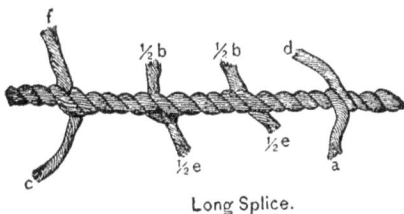

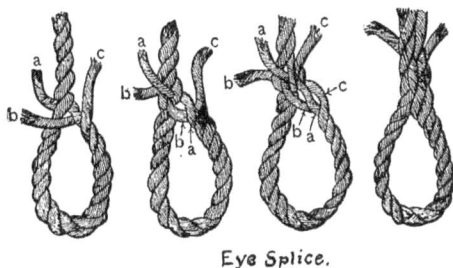

FIGURE 128.—Splices.

FIGURE 129.—Square lashing.

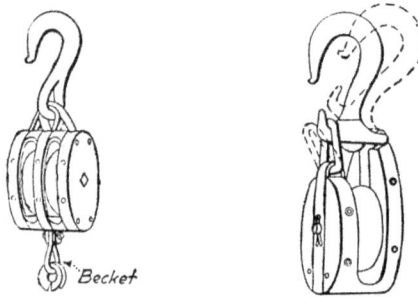

FIGURE 130.—Double wooden block and snatch block.

FIGURE 131.—Lashing for tripod.

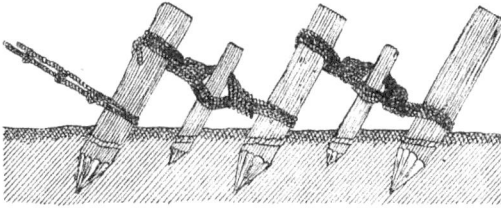

① Holdfasts.

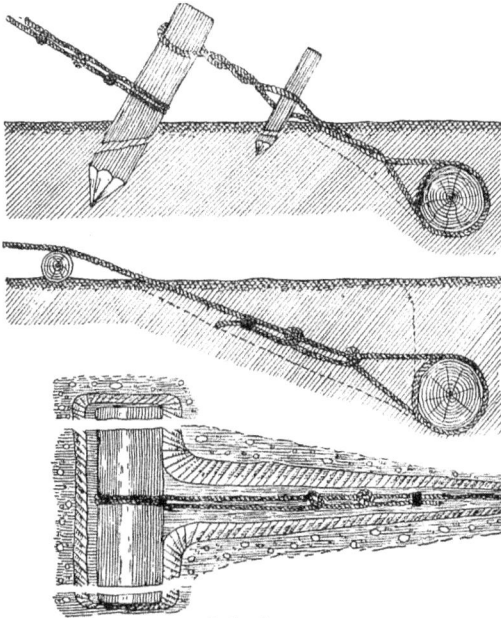

② Deadman.

FIGURE 132.—Anchorages.

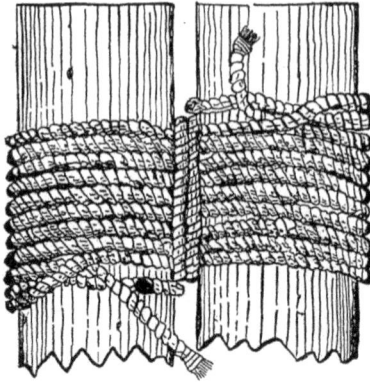

FIGURE 133.—Lashing for shears.

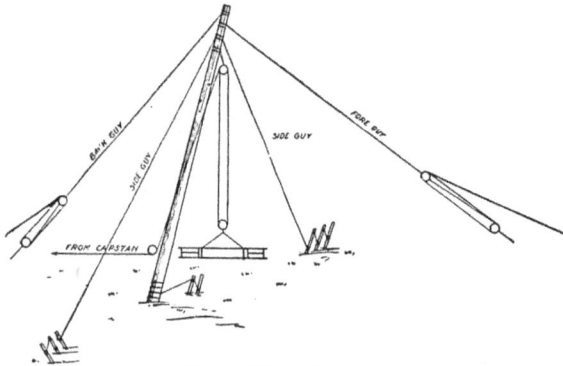

FIGURE 134.—Gin pole.

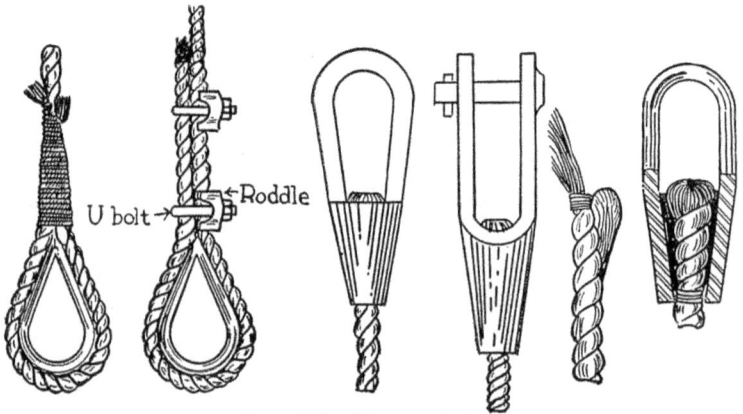

FIGURE 135.—Wire rope fittings.

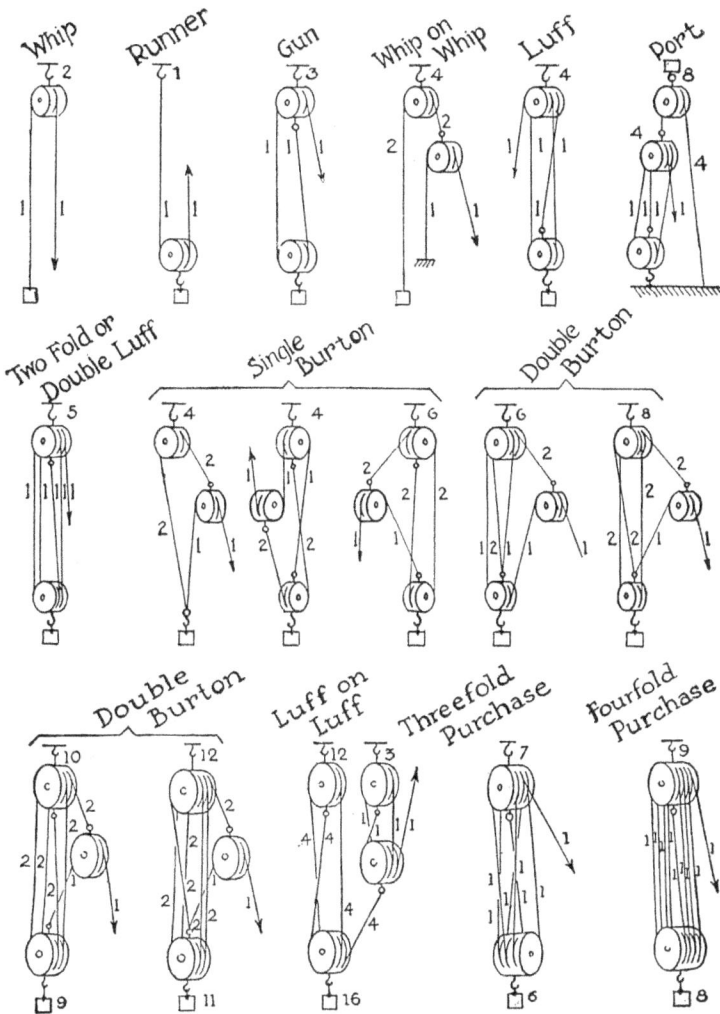

FIGURE 136.—Block and tackle combinations.

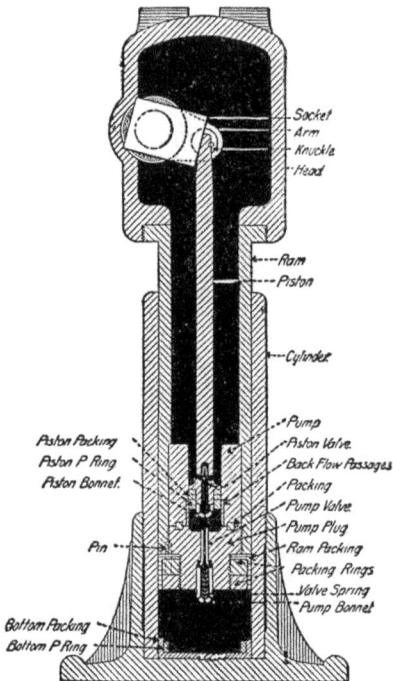

FIGURE 137.—Base jack.

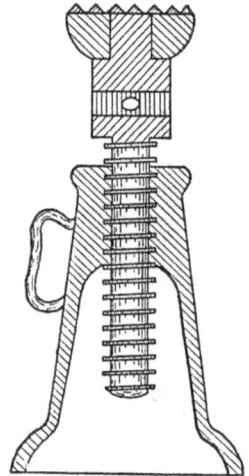

FIGURE 138.—Screw jack.

Lowering valve
Knuckle
Pump nut
Pump
Pump valve
Pump valve spring

Ram packing
RP Ring
R.P Ring nut
Bottom packing
B.P Ring
Pump small nut

Ram head
Ram
Cylinder

Cap
Socket
Knuckle
Reservoir
Piston
Pump
Piston valve
Piston packing
Piston packing ring
Piston valve bonnet

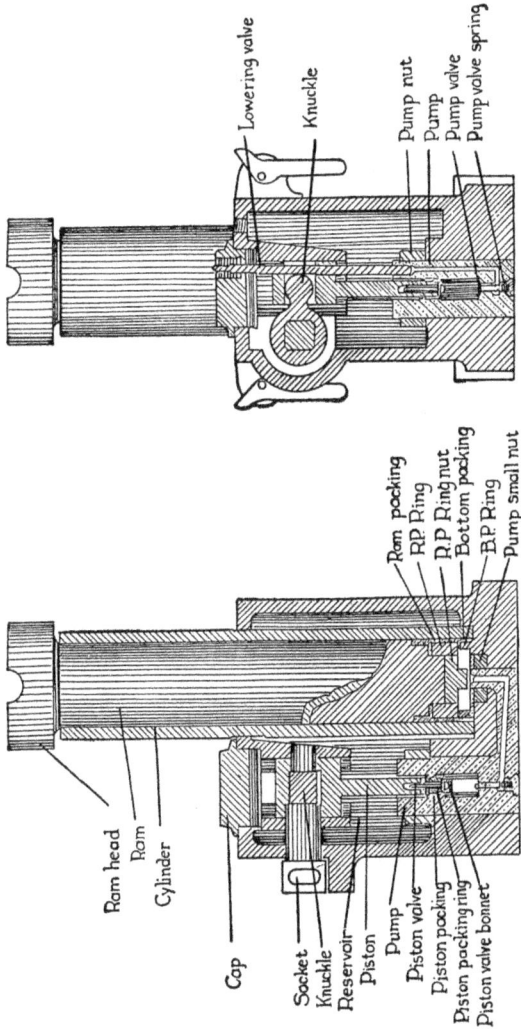

FIGURE 139.—Horizontal hydraulic jack.

291

Section III

MAP READING

94. Scales, contours, and conventional signs.—*Q*. What is a map? *A*. A map is a picture of an area of ground, which shows certain important features accurately to scale.

Q. Do the features shown on a map appear as they do on the ground? *A*. No. They are represented by symbols called conventional signs, which resemble the actual features as nearly as practicable.

Q. What is a topographical map? *A*. One which (according to its scale) shows all the natural and artificial features of the terrain, such as hills, valleys, streams, woods, roads, towns, houses, bridges.

Q. What is a military map? *A*. A military map is one which shows particularly those features and conveys that information, which are important for military purposes.

Q. What is meant by map reading? *A*. Map reading is the art of understanding the information given by the map.

Q. What is meant by the scale of a map? *A*. The scale of a map is the relation between any distance shown on the map and the corresponding distance on the ground. It is always the same for any one map.

Q. How is the scale of a map indicated or expressed? *A*.

(1) As a representative fraction (RF), such as 1/5,000 (or 1 : 5,000), which means that any distance on the map is 1/5,000 of the corresponding distance on the ground. The RF is always expressed with a numerator of unity. The RF is always the ratio of a map distance to the actual or ground distance it represents. Thus if 2 inches (on the map) represent 10 miles on the ground, we can at once express the relation thus:

$$\frac{\text{Map distance}}{\text{Ground distance}} = \frac{2 \text{ inches}}{10 \text{ miles}}$$

We must then reduce the numerator and denominator to the same unit, and then reduce the numerator to unity to get the RF, as follows:

$$\frac{2 \text{ inches}}{10 \text{ miles}} = \frac{2 \text{ inches}}{10 \text{ miles} \times 5{,}280 \text{ feet} \times 12 \text{ inches}} =$$

$$\frac{2 \text{ inches}}{633{,}600 \text{ inches}} = \frac{1}{316{,}800} = \text{RF}$$

(2) In words and figures, such as 6 inches=1 mile, meaning that 6 inches on the map represent 1 mile on the ground. From this we can easily get the RF, thus:

$$\frac{6 \text{ inches}}{1 \text{ mile}} = \frac{6 \text{ inches}}{1 \text{ mile} \times 5{,}280 \text{ feet} \times 12 \text{ inches}} =$$

$$\frac{6 \text{ inches}}{63{,}360 \text{ inches}} = \frac{1}{10{,}560} = RF$$

(3) By a graphical scale drawn on the map, which shows ground distances in their usual units, such as miles, thousands of yards, or hundreds of feet, as they appear on the map. A graphical scale is easily made if we know the RF of the map. Thus suppose the RF is 1:5,000 and we wish a graphical scale to read to 1,000 yards. Since any distance on the map is 1/5,000 of the same distance on the ground, 1,000 yards on the map will be as follows:

$$\frac{1{,}000 \text{ yards}}{5{,}000} = \frac{1 \text{ yard}}{5} = 0.2 \text{ yard} = 0.2 \text{ yard} \times 36 \text{ inches} = 7.2 \text{ inches}$$

that is, one 1,000-yard division of our scale will be 7.2 inches long. This can be divided into 10 equal parts, each of which will represent 100 yards. (See fig. 143 for examples of graphical scales.)

Q. Why do we have maps of different scales? *A.* The scale of a map must be large enough to show the particular features about which we need information. Thus, a map showing the positions of all buildings and streets in a town must be of much larger scale than a map intended only to show the size and positions of the various counties in a state. A small scale map is one which shows a large area in a small space. Thus 1 inch equals 100 miles would be a very small scale map on which very little detail could be shown. A map on a scale of 1 inch equals 25 feet would be a very large scale map on which individual trees could be shown in their exact positions. The first map could be used by a general planning a large campaign, the second by an architect laying out a plan of a house and grounds.

Q. What determines the proper scale of a map? *A.* It should be just large enough to show the detail necessary to serve the purpose for which it is to be used.

Q. How do you determine direction on a map? *A.* By referring to an arrow on the map which points due north. It is called the meridian.

Q. How is direction measured and indicated on a map? *A.* By azimuth as in gunnery. Azimuth on a map is measured from the north point of the meridian, clockwise around the horizon. It is measured in degrees, minutes, and seconds, or in mils.

Q. What are the cardinal points of direction, and what are their azimuths? *A.* North, east, south, and west. Moving clockwise around the horizon; north, the origin, is azimuth 0°; east is azimuth 90°; south is azimuth 180°; west is azimuth 270°.

Q. What is meant by orienting a map? *A.* Placing the map in such a position that the meridian or arrow on the map points to north on the ground. Every line on the map will then be parallel to the line on the ground which it represents, and all the features on the map will be in the same relative positions as the actual objects on the ground.

Q. What is elevation? *A.* The elevation of any point is its vertical height in feet above some level, usually sea level.

Q. How is elevation indicated on a map on a flat piece of paper? *A.* By means of contours.

Q. What is a contour? *A.* An irregular line joining all points at the same elevation. A contour is thus a level or horizontal line.

Q. How are contours separated vertically? *A.* By some definite interval, such as 5, 10, 20, 50, or 100 feet, depending on the scale of the map and the nature of the ground. This constant interval is known as the vertical interval or *contour interval* of the map. Certain contours are numbered with their height (in feet) above sea level.

Q. Do contours show the ground forms, such as hills, valleys, and ridges? *A.* Yes. When one has become familiar with them they show accurately all the forms of nature.

Q. Mention briefly the principal characteristics of contours. *A.*

(1) A contour is a horizontal line joining points of equal elevation.

(2) Contours are spread at uniform vertical intervals.

(3) Every contour is a continuous closed curve. (It may not close within the limits of the map.)

(4) There may be any number of separate contours of the same elevation.

(5) A small, closed contour indicates either a hilltop or a depression.

(6) Contours never touch or cross each other except in the case of a vertical or overhanging cliff.

(7) Contours are at right angles to the lines of steepest slope.

(8) The horizontal spacing of contours indicates the degree of slope, steep if they are close together, gentle if they are far apart. They also indicate the kind of slope, uniform, concave, or convex.

(9) Valley contours are usually of V-shape, and hill or ridge contours of U-shape.

(10) Adjacent contours resemble each other.

Q. How is the elevation of a point between two contours determined? *A.* By its relative distance from the contours on either side. Thus in figure 140, the elevation of the number "46" is about 527 feet.

Q. Point out the characteristic ground forms in figure 140 and explain how they are shown by the contours. *A.* See figure 140.

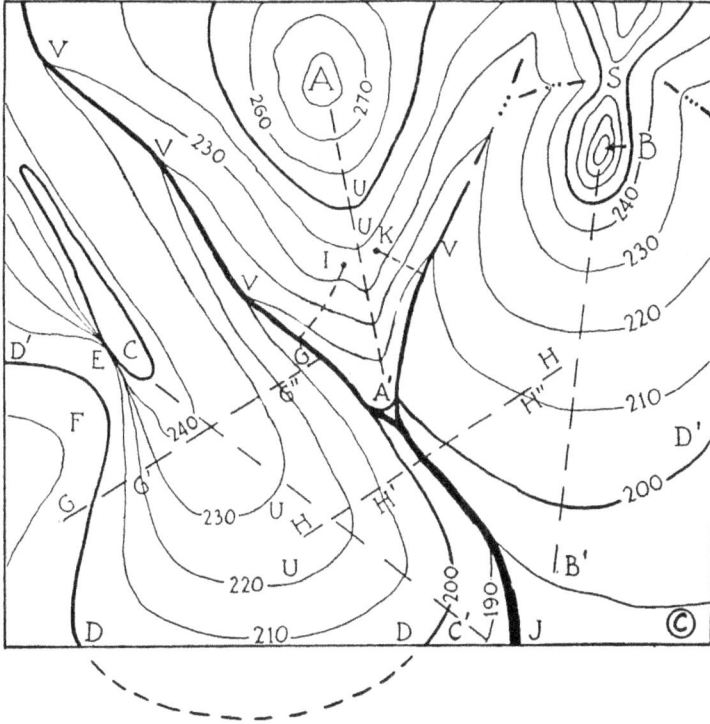

A. B. Hilltops or peaks.
C. Ridge.
DD. Contours close on themselves.
E. Cliff.
AA'. Uniform slope.
BB'. Concave slope.
CC'. Convex slope.

IG. Line of steepest slope.
B. Steep slope.
B'. Gentle slope.
S. Saddle.
UU. Hill contours.
VV. Valley contours.
A'. Stream junction.

FIGURE 140.—Typical ground forms as shown by contours.

Q. What is a slope? *A.* The inclination of ground to the horizontal. The slope of a road is called its grade.

Q. How can you determine the average grade of a length of road from a contoured map? *A.* Measure the horizontal distance along the road in feet, using the scale of the map. Find from the contours the

difference in elevation between the two ends. The difference in eleva-
vation of the two ends, divided by the length of the road (both in feet),
will give the average grade in percent.

Q. How is a steep grade indicated on a map? *A.* By contours close
together, showing a considerable change of elevation in a short distance.

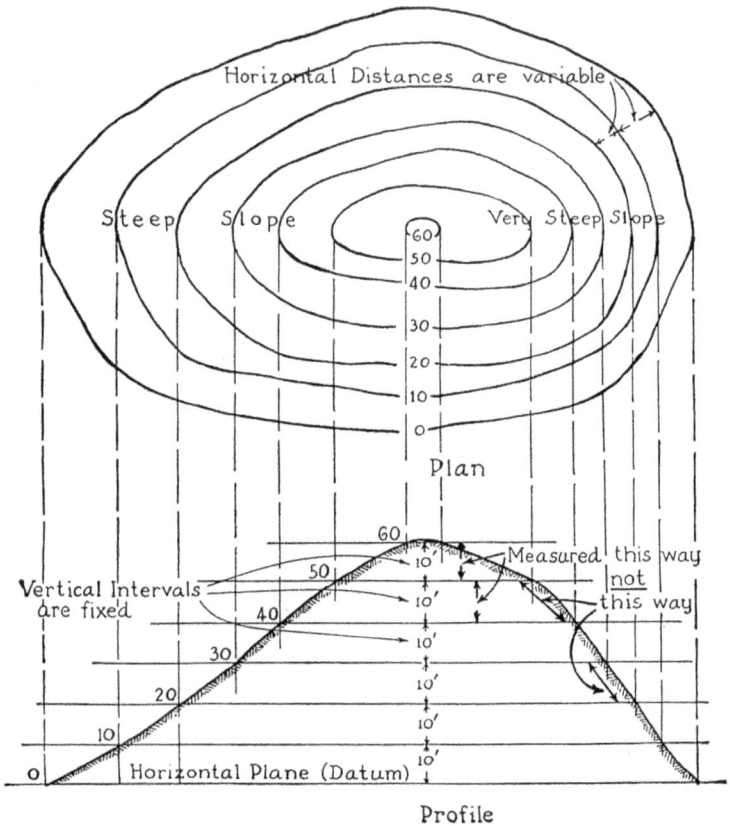

FIGURE 141.—Plan and profile of hill.   (Contours are lines that would be left on the ground
by successive horizontal slices, 10 feet thick.)

Q. What is a profile? *A.* It is a section of the ground as it would
appear if it were sliced vertically with a huge knife.

Q. Give a simple method of making a profile. *A.* To make a profile
of the hill shown in figure 141, along the line of the words "Steep slope,
Very steep slope"; draw a number of parallel lines at uniform inter-

vals, as in the lower figure, numbering them at 10-foot intervals from 0 to 60 feet (the limiting elevations of the hill). Mark each point where a contour cuts the line "Steep slope," and project these points vertically down to the correspondingly numbered lines. Join the points thus found by a line. It will be the contour of the hill on the line "Steep slope."

Q. What are conventional signs? *A.* They are the symbols used by map makers to show the various features of the terrain. As nearly as practicable they resemble or suggest the features they are intended to show.

Q. Point out and name the conventional signs shown in figure 142. *A.* See figure 142.

Q. What colors are used on standard topographical maps and what do these colors mean? *A.* Colors are used to show certain classes of features. On the standard topographical map they show the following:

(1) *Black.*—All artificial features, such as houses and roads.

(2) *Blue.*—All water, such as streams, ponds, lakes.

(3) *Green.*—Vegetation, such as woods and grassland.

(4) *Brown.*—The ground forms as shown by contours.

95. Location of position by coordinates.—*Q.* How are military maps divided? *A.* Into squares 1,000 yards on a side by two sets of parallel lines, one of these sets being at right angles to the other (figure 143).

Q. What is this system of lines or squares called? *A.* A grid system, or system of rectangular coordinates.

Q. What is the use of this grid or coordinate system? *A.* It is used to make it possible to describe and locate points on a map by referring their positions to the coordinate lines.

Q. What are the coordinate lines running from left to right called? *A.* X-lines.

Q. What are the other coordinate lines called? *A.* Y-lines.

Q. Describe the grid system for the continental United States. *A.* The country is divided into zones running north and south, each covering 9° latitude. The central true meridians of adjacent zones are 8° apart, hence there is an overlap of 1° at the boundaries, included in both adjacent zones. The west longitudes of the central meridians of the zones are 73°, 81°, 89°, 97°, 105°, 113°, and 121°. In each zone the Y-lines are all parallel to the central true meridian. The direction of the Y-line at any point is called grid north; it is the same as true north only at the center of the zone. The X-lines are perpendicular to the Y-lines, and so they are true east and west lines only at the center

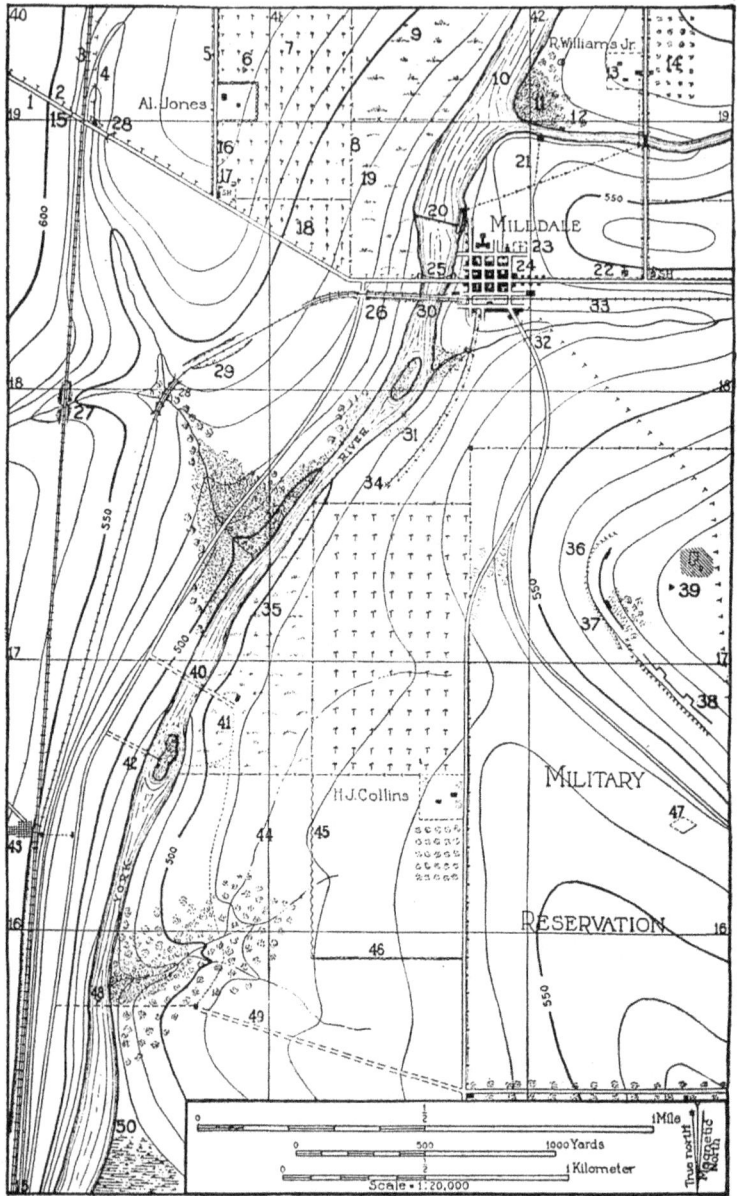

FIGURE 142.—Conventional signs.

Numerical key

1. Good motor road.
2. Telephone or telegraph line.
3. Double track standard gage railroad.
4. Stream or creek (blue on a four-color map).
5. Fence, smooth wire.
6. Triangulation point or primary traverse station.
7. Cornfield.
8. Fence, barbed wire.
9. Tall tropical grass.
10. River (blue on a four-color map).
11. Woodland (deciduous trees).
12. Lone trees.
13. Buildings in general.
14. Orchard.
15. Railroad crossing, railroad beneath.
16. Fence of any kind (or board fence).
17. Schoolhouse.
18. Cultivated field, sugarcane.
19. Grassland in general.
20. Dam.
21. Electric power transmission line.
22. Church.
23. Cemetery.
24. City, town, or village.
25. Bridge, suspension.
26. Railroad crossing, railroad above.
27. Fill.
28. Bridge, steel.
29. Cut, railroad.
30. Bridge, truss, for standard gage railroad.
31. Narrow-gage railroad.
32. Bridge, highway.
33. Railroad, single track, standard gage.
34. Mine or quarry of any kind (or open cut).
35. Pasture or grassland in general.
36. Wire entanglement.
37. Low or concealed entanglement.
38. Trenches (dotted when proposed).
39. Demolitions.
40. Ford, general symbol for vehicle ford.
41. Good pack trail or footpath.
42. Bridge, foot.
43. City, town, or village (generalized).
44. Intermittent stream.
45. Worm fence.
46. Stone fence.
47. Tank trap.
48. Equestrian ford.
49. Road, poor motor or private.
50. Marsh in general.

Alphabetical key

of the zone. In each zone the origin of coordinates, or zero point, is to the west and south of the zone, and hence all coordinates are positive. (See fig. 143.)

Q. What is the *X*-distance or *X*-coordinate of a point on a map? *A.* The distance of that point to the right of the origin of measurement measured along an *X*-line.

Q. What is the *Y*-distance or *Y*-coordinate of a point on a map? *A.* The distance of that point upward from the origin of measurement measured along a *Y*-line.

Q. What is the origin or point from which measurements are made? *A.* A point off the map to the west and south. The distance from this origin, in thousands of yards, is indicated on each *X*-line and each *Y*-line.

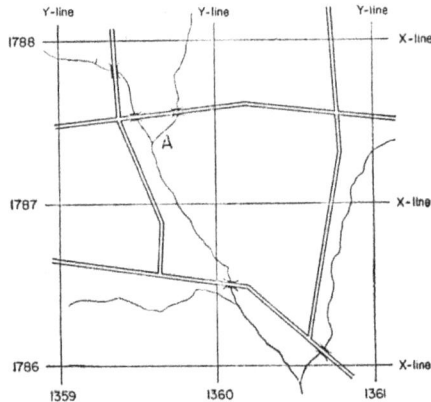

FIGURE 143.—Rectangular coordinates. 1,000-yard grid printed upon a terrain map (1 : 20,000 scale). (Not reproduced to scale.)

Q. In order to locate a point on a map, what information is useful? *A.* The *X*- and *Y*-coordinates of that point.

Q. In what order are the coordinates of a point always given? *A.* First the *X*-coordinate and then the *Y*-coordinate.

Q. How can their order be remembered easily? *A.* Remember that in the alphabet *X* comes before *Y*, or remember the rule READ RIGHT UP.

Q. How can one *X*-line be distinguished from the other *X*-lines or one *Y*-line from the other Y-lines? *A.* Each set of lines is numbered.

Q. How are the *X*-lines numbered? *A.* In the left- and right-hand margins, increasing upwards from the bottom of the map.

Q. How are the *Y*-lines numbered? *A.* Along the upper and lower margins, increasing from left to right of the map.

Q. Is the number of a coordinate line given in the margin of a map the full number of that line? *A.* No. It is only a part of its full number.

Q. Are there any coordinate lines on a map that have their full numbers given? *A.* Yes. The first *X*-line at the bottom of the map and the first *Y*-line at the left of the map.

Q. Explain in detail how to read the coordinates of any point on a gridded map. *A.* The coordinates of any square in the grid system are the *X*- and *Y*-coordinates of the lower left-hand (southwest) corner of the square. The coordinates of the lower left-hand square in figure 144 are (1,364,000–1,790,000) but are usually written (1364–1790), it being understood that each number represents thousands of

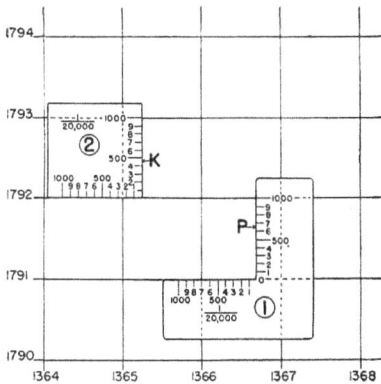

Figure 144.—To plot position of point with coordinate scale on terrain map (1 : 20,000 scale) having 1,000-yard grid. (Not reproduced to scale.)

yards. These coordinates also locate to the nearest 1,000 yards any point in the lower left-hand square of figure 144. To locate a point more closely we can assume that each side of the square in question is divided into 10 equal parts, each of which represent 100 yards. Then the coordinates of the center of the lower left-hand square would be expressed as (1364.5–1790.5), which locates the center to the nearest 100 yards in each direction. The method may be further refined to read to the nearest 10 yards.

Q. What is a coordinate scale? Explain its use. *A.* A coordinate scale is a right-angled ruler made of thin metal, celluloid, or other material with scales on it equal in length to the grid interval of the map being used (fig. 143). To plot any point on the map, for instance the point P of which the coordinates are given as (66.70–91.65), place the coordinate scale on the map in position 1 as shown in figure 144. The position of P can be marked at once with a pin or sharp pencil. It should be noted here that the first two numbers of the coordinate expression (1366.70–1791.65) have been dropped because they are common to all points in the section of map under consideration. The process of reading the coordinates of a point appearing on the map is the reverse of the method given for plotting a point on the map.

Q. How many norths are indicated on an artillery map? A. Three. True north, magnetic north, and grid north (fig. 143).

Q. Define each. A.

(1) True north is geographic north, or the direction of the North Pole.

(2) Magnetic north is the direction in which the compass points when used in the area covered by the map. The angular difference between true north and magnetic north is called the magnetic declination.

(3) Grid north is the direction in which the Y-lines of the coordinate system point.

Q. When indicating direction which north is generally used? A. Grid north.

Q. On what kind of maps is such a system of coordinates usually found? A. On topographical maps.

Q. What type of maps are often used to cover harbors and water areas? A. Coast and Geodetic Survey maps.

Q. Do these maps generally have a system of grid coordinates? A. No. There is, however, a local system of grid coordinates placed on these maps so points can be located in the same way.

Q. What do these maps show? A. Coast line, channels, depths of water, location of lights, channel markers, and certain permanent features of the terrain along the shore.

(In addition the candidate should be required to give the coordinates of certain points on a map.)

96. Following route indicated on map.—Q. How is a route, selected from a map, usually indicated? A. It is indicated by naming successive points along the route that can easily be identified locally.

Q. How are the best roads identified on the ground? A. Usually they will be paved, and will be wider, straighter, and have easier grades than secondary roads. If not paved they will at least be wide and show signs of traffic. They usually have telegraph or telephone lines running parallel to them. Except in very sparsely settled country the principal roads will also be indicated by signposts at intersections and Federal and State numbers, which are shown on commercial route maps.

Q. What points along a route are most easily identified? *A.*

(1) Large towns.

(2) Villages.

(3) Important crossroads or junctions.

(4) Crossings of large streams.

(5) Railroad crossings.

(6) Crests of important hills or ridges.

(7) Passes or gaps through lines of hills.

Q. How may an indicated route be traced on a map? *A.* By locating in succession the towns, crossroads, and other important points named.

Q. How are crossroads and road junctions indicated on a map? *A.* By numbers sometimes followed by a letter as 423 or 423–A.

Q. How would crossroad 418–A be written in indicating a route? Road junction 403? *A.* CR 418–A. RJ 403.

Q. To what do the numbers in the previous answer refer? *A.* To the elevation of the cross road or road junction.

Q. In case a cross road or junction is not plainly marked on the map, how would it be indicated? *A.* By its coordinates as CR 4365.5–6427.3) or RJ (7295.4–8665.4).

(The candidate should be able to trace out on a map a route that has been indicated to him.)

97. Data as to roads, bridges, fords, grades, and swamps.— *Q.* What information concerning a road can usually be obtained from a good topographical map? *A.*

(1) The distance between any two points.

(2) Whether or not the road is paved, and often the kind of pavement.

(3) The width of the road, that is whether narrow, wide, or quite wide.

(4) The steepness and length of important grades.

(5) The stream crossings, whether bridges, fords, or ferries, and sometimes the kind and principal dimensions of bridges; width, depth, nature of bottom, and velocity of current in the case of fords; and the kind of ferry.

Q. What is usually the most critical question in the selection of a route? *A.* The stream crossings.

Q. How would you decide the question as to whether a certain bridge was safe for the transport accompanying your organization? *A.* Reports on the practicability of all bridges should be secured in advance if possible. If not, the following observations will indicate the safety of bridges in most cases:

(1) Bridges on important routes habitually carry heavy commercial trucks and busses, moving at high speed, and are therefore safe for artillery transport moving slowly.

(2) If a bridge is massive, and reasonably new, or apparently in good condition as to flooring, paint, etc., it is probably safe.

(3) Bridges may be compared with similar bridges that have been crossed. If they look too light or appear to be of older design than other bridges, they should be regarded with suspicion.

(4) If there is any chance that the enemy may have tampered with a bridge, its abutments, piers, flooring, and truss members or cables should be examined to make sure they are intact.

(5) If in doubt about any bridge, send across some lighter vehicle and watch the bridge as it crosses. If there is no excessive sway or vibration the bridge is probably safe for the next heavier load. Send loads across one at a time and at very slow speed.

Q. How may the practicability of a ford be determined? *A.* Note the swiftness of the current. Send a line of men to wade across, preferably barefoot, to determine the depths, nature of the bottom and whether the banks are steep or slippery or both. Some of the lighter vehicles may then be sent across, and these can assist the heavier vehicles by pulling them out on the far side if necessary.

Q. In case a bridge or ford proves impassible what should be done? *A.* Detour to another crossing.

Q. What in general can you say as to the practicability of routes? *A.* That in general important main routes are practicable for artillery transport; that in the case of less important routes it is desirable to have a reconnaissance made in advance by competent experts; that any route that lacks a bridge at an important crossing is of doubtful practicability.

(The candidate should be required to examine a route shown on a map, to give all the information concerning it that can be obtained from the map, and his opinion as to the practicability of the route.)

SECTION IV

INDICATION, IDENTIFICATION, AND CHARACTERISTIC FEATURES OF CLASSES OF AIRCRAFT

98. Classes and types of aircraft.—*Q.* What are the two general classes of aircraft? *A.* Heavier-than-air and lighter-than-air.

Q. Name the general types of lighter-than-air aircraft. *A.* Observation balloons and dirigible airships.

Q. What are the general types of airships? *A.* Nonrigid, semirigid, and rigid.

Q. What are the heavier-than-air aircraft generally called? *A.* Airplanes or aeroplanes, seaplanes, flying boats, amphibians.

Q. What are the general types of combat airplanes used by the United States Army? *A.*

(1) Pursuit.

(2) Bombardment.

(3) Reconnaissance, observation, and liaison.

(4) Transport.

Q. How are pursuit airplanes classified? *A.*

(1) Interceptor.

(2) Single-place fighter.

(3) Multiplace fighter.

Q. How are bombardment airplanes classified? *A.*

(1) Heavy.

(2) Medium.

(3) Light.

99. Missions of aircraft.—*Q.* What is the normal mission of pursuit airplanes? *A.* The interception, attack, and destruction of enemy aircraft in the air. The interceptor is usually a single seater with one or two powerful engines. The single-place fighter is used for escort and patrol in addition to normal pursuit missions. The multiplace fighter is used for escort and patrol duty near important objectives and against ground-troop formations.

Q. What are the normal missions of heavy and medium bombardment airplanes? *A.* To carry heavy bomb loads to great distances for attack of material objectives, and also to conduct long range strategic reconnaissance over land and sea.

Q. What are the normal missions of light bombardment airplanes? *A.* Light bombardment airplanes (formerly designated as attack) are designed to attack objectives of light construction, routes of communication, airdromes, troop movements, and concentrations of troops in the open or under light shelter. The light bombardment airplane is the striking element of combat aviation which operates in direct support of ground forces. Identification of this type of airplane is especially important to ground troops.

Q. What are the normal missions of reconnaissance, observation, and liaison airplanes? *A.* They gather information of the enemy. The two latter types operate in conjunction with our own forces performing fire-adjustment missions for artillery, maintaining contact with our front lines and marching columns, and carrying on other command, liaison, and courier missions.

Q. What are the missions of transport airplanes? *A.* Transport airplanes are not strictly a combat type of airplane. They are used for the transportation of personnel and supplies. Their importance is rapidly increasing when we consider the transportation of air landing troops, parachute troops, and important supplies.

100. Naval aircraft.—*Q.* What types of airplanes are employed by our Navy, and to what types of Army airplanes do they correspond? *A.*

(1) Scouting-observation airplanes corresponding to observation airplanes.

(2) Fighter airplanes corresponding to pursuit airplanes.

(3) Torpedo-bombardment airplanes corresponding to bombardment airplanes.

(4) Patrol airplanes which do not correspond to any special type of Army airplane. The Navy has no type of airplane corresponding to the Army light bombardment airplane.

Q. Does the Navy make more extensive use of the biplane type of airplane than the Army? *A.* Yes. They are used on carriers and on board other types of warships being launched from catapults. They are used for this purpose because they are more stable in flight at low air speeds than monoplanes.

Q. What are seaplanes and flying boats? *A.* They are airplanes equipped with floats (pontoons) or boat-shaped hulls instead of wheels, so that they may alight on water. Seaplanes have floats while flying boats have hulls.

Q. What is an amphibian airplane? *A.* It is an airplane having a boat-shaped hull, and also equipped with wheels (that can be pulled up when operating on water) so that it can alight or take off from either a land or water surface.

101. Identification and indication of aircraft.—*Q.* Why is it important that ground personnel be familiar with the appearance in flight, method of operation, and characteristic sounds of airplanes? *A.* These factors are the means by which airplanes are identified and indicated.

Q. What are the basic flight positions used for ready recognition of airplane types? *A.*

(1) Coming flight or front view.

(2) Passing flight or side view.

(3) Flight at lower altitude or top view.

(4) Overhead flight or bottom view.

(5) Maneuvering flight or perspective view.

Q. What is meant by coming flight or front view? *A.* All positions of flight in which only a general head-on view of the airplane may be had.

Q. What is meant by passing flight or side view? *A.* All positions of flight in which the side of the fuselage, vertical fin, and rudder are the major surfaces presented to view.

Q. What is meant by flight at lower altitude or top view? *A.* All positions of flight in which the upper sides of wings, fuselage, and horizontal tail surfaces are the major surfaces presented to view.

Q. What is meant by overhead flight or bottom view? *A.* All flight positions in which the under sides of wings, fuselage, and horizontal tail surfaces are presented to view.

Q. What is meant by maneuvering flight or perspective view? *A.* All flight positions which are different from straight and level flight. It includes banking, turning, climbing, diving, and combinations of such maneuvers. The airplane may present, momentairly at least, nearly all of the views presented under other conditions of flight.

Q. What characteristics of outline of the airplane are most readily seen in overhead flight? *A.*

(1) *Shape of wing.*—The general shape and proportion of wings, as long and narrow, short and stubby.

(2) *Type and shape of nose.*—Nose extends much or little in advance of leading edge of wings; that is, plane is long-nosed or short-nosed.

(3) *Length and shape of fuselage.*—Compare the relatively short fuselage of the small and medium sized airplanes with the long, slender, streamlined appearance of the larger types.

(4) *Location and number of engines.*—In single-engined airplanes the engine is located in the nose and by its type determines the shape of the nose; that is, with radial engines the nose is blunt and stubby, while with in-line and V-type engines the nose is more slender and pointed. In multiengined airplanes the engines are usually housed in nacelles extending from the leading edge of the wings. In the unusual pusher types, the engines extend from the trailing edge of the wings. Even at great altitudes when the number of engine nacelles cannot be exactly determined, their presence will give an unmistakable irregular outline to the wings warranting identification as multiengined.

Q. What characteristics of outline are most readily seen in passing flight? *A*.

(1) *Shape and outline of fuselage.*—It is short and chunky in smaller pursuit types; elongated and streamlined in larger types; long and thick bodied in larger bombardment types. Note outline being broken by such parts as cockpits, canopies over cockpits, and gun turrets.

(2) *Shape of nose.*—It may be slender and pointed, blunt and stubby, smoothly rounded, or shark-nosed.

(3) *Size of fin and rudder.*—Note the relative size of the vertical fin and rudder compared to the fuselage.

Q. What characteristics of outline are most readily seen in coming or going flight? *A*.

(1) *Relationship of wings to fuselage.*—Has high-wing, mid-wing, low-wing, or parasol-wing type; dihedral angle, pronounced, moderate, or practically zero.

(2) *Number of engines.*—The irregularity of outline of wings will indicate a multiengined type.

(3) *Features of vertical tail members.*—It is usually possible to identify single- and double-rudder types.

(4) *Undercarriages.*—Nonretractable landing gear is usually plainly visible.

Q. What characteristics of outline are most readily seen in maneuvering flight? *A*. All the features previously pointed out may be momentarily visible.

Q. What characteristic methods of operation of pursuit assist in its identification? *A*. Pursuit normally operates in formation with the squadron of 18 airplanes as the largest group operating as a unit. An observer noting one such formation should look below and to the front of it and above and to the rear of it for other units.

Q. What characteristic methods of operation of heavy and medium bombardment assist in its identification? *A*. They operate in column of three-plane elements (route column) with successive elements stepped up or down from front to rear. They usually fly straight courses at medium or high altitude unless attacked from the air or by antiaircraft fire.

Q. What characteristic methods of operation assist in the identification of light bombardment? *A*. They operate in formation at minimum or medium altitudes. They use the three-plane element echeloned to the rear at approximately the same altitude. The normal operating unit is the squadron of nine airplanes; the largest formation is the group of three squadrons. This type of aviation supports the operations of ground troops.

Q. What characteristics of operation of reconnaissance airplanes assist in their identification? *A.* They operate at any altitude from low to high; usually operate singly; fly straight courses unless attacked. Bombardment airplanes may perform long-range reconnaissance.

Q. What characteristic methods of operation assist in identification of observation and liaison airplanes? *A.* They operate almost entirely within own lines; fly singly on various courses at low and medium altitudes; will be seen circling over own troops and troop columns to drop messages and observe panels.

Q. What are some of the characteristic sounds of pursuit airplanes in flight? *A.* Pursuit airplanes in flight are characterized by sounds of fast rhythm, high pitch, moderate volume, and by extreme variations in pitch and tone while maneuvering.

Q. What are some of the characteristic sounds of heavy and medium bombardment airplanes while in flight? *A.* They have a fairly deep pitch, a moderately heavy volume, and a steady tone and rhythm.

Q. What are some of the characteristic sounds of light bombardment airplanes while in flight? *A.* They have a heavy volume of sound due to low altitude; a fairly deep pitch, with tone and rhythm steady but varying considerably when maneuvering.

Q. State, in the order in which given, what information is given and the terms used in indicating aircraft during daylight. *A.*

(1) Designation of the reporting station by name or number.

(2) Number of airplanes, when they can be counted. If they cannot be counted the words "several" or "many" may be used.

(3) Type of airplane, such as observation, pursuit, etc., when they can be identified. In other cases the word "airplane" is used.

(4) Altitude, in general terms as follows: very low (below 500 yards); low (500 to 2,000 yards); medium (2,000 to 5,000 yards); or high (over 5,000 yards).

(5) Location, by the sector in which or toward which the aircraft are flying.

(6) Direction of flight, by one of the eight points of the compass: north, NE, east, SE, south, SW, west, NW.

Q. State which of these elements of information are given in indicating aircraft at night. *A.* Designation of reporting station, number of airplanes (one, several, or many), altitude, and location.

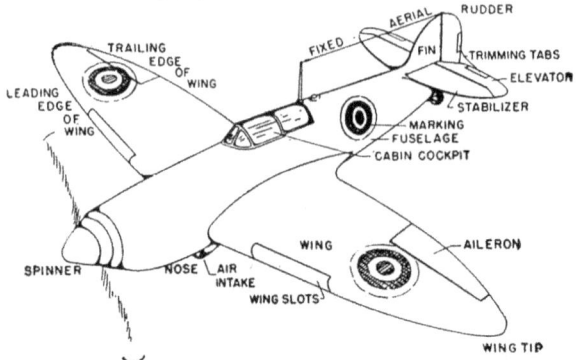

FIGURE 145.—Nomenclature of airplane parts.

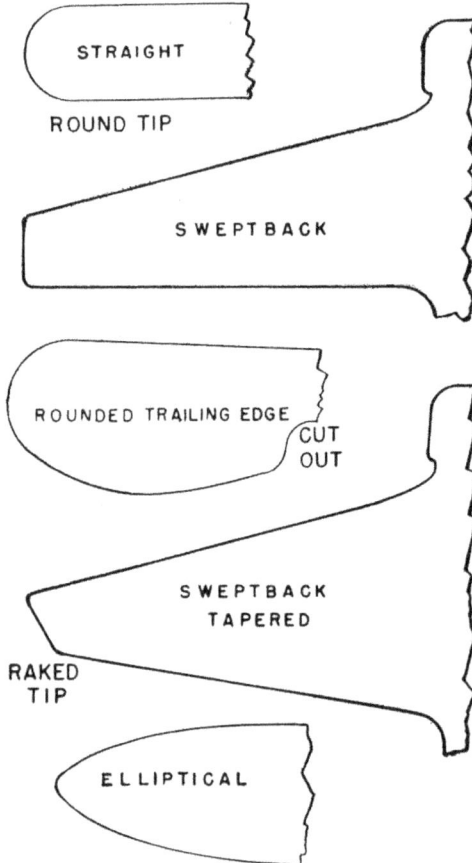

FIGURE 146.—Wing shapes.

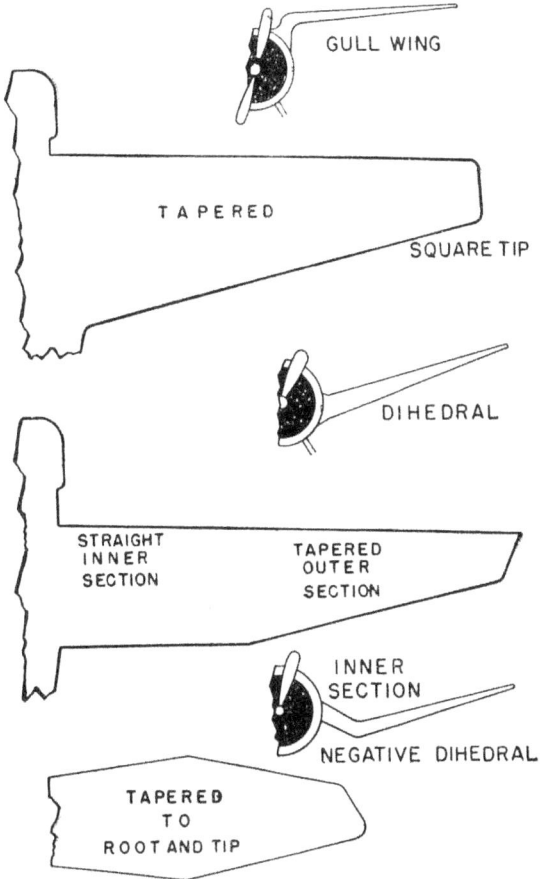

GULL WING

TAPERED

SQUARE TIP

DIHEDRAL

STRAIGHT
INNER
SECTION

TAPERED
OUTER
SECTION

INNER
SECTION

NEGATIVE DIHEDRAL

TAPERED
TO
ROOT AND TIP

FIGURE 146.—Wing shapes—Continued.

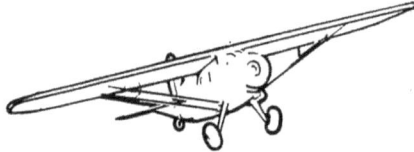

HIGH-WING

MIDWING

LOW-WING

(PARASOL MONOPLANE)

FIGURE 147.—Monoplanes.

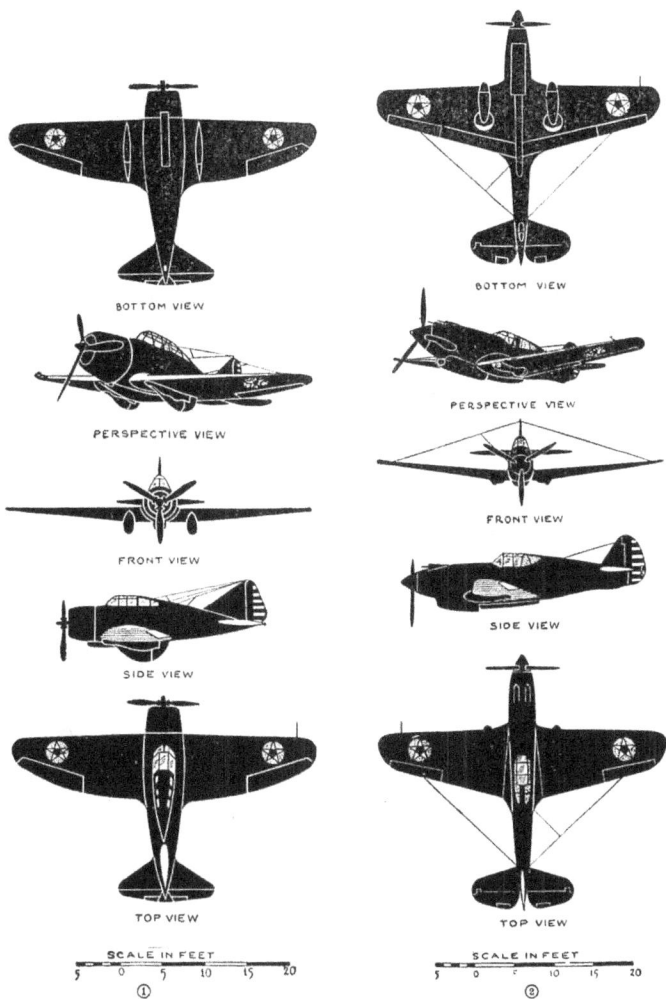

BOTTOM VIEW

PERSPECTIVE VIEW

FRONT VIEW

SIDE VIEW

TOP VIEW

SCALE IN FEET

①

BOTTOM VIEW

PERSPECTIVE VIEW

FRONT VIEW

SIDE VIEW

TOP VIEW

SCALE IN FEET

②

FIGURE 148.—Pursuit airplanes.

313

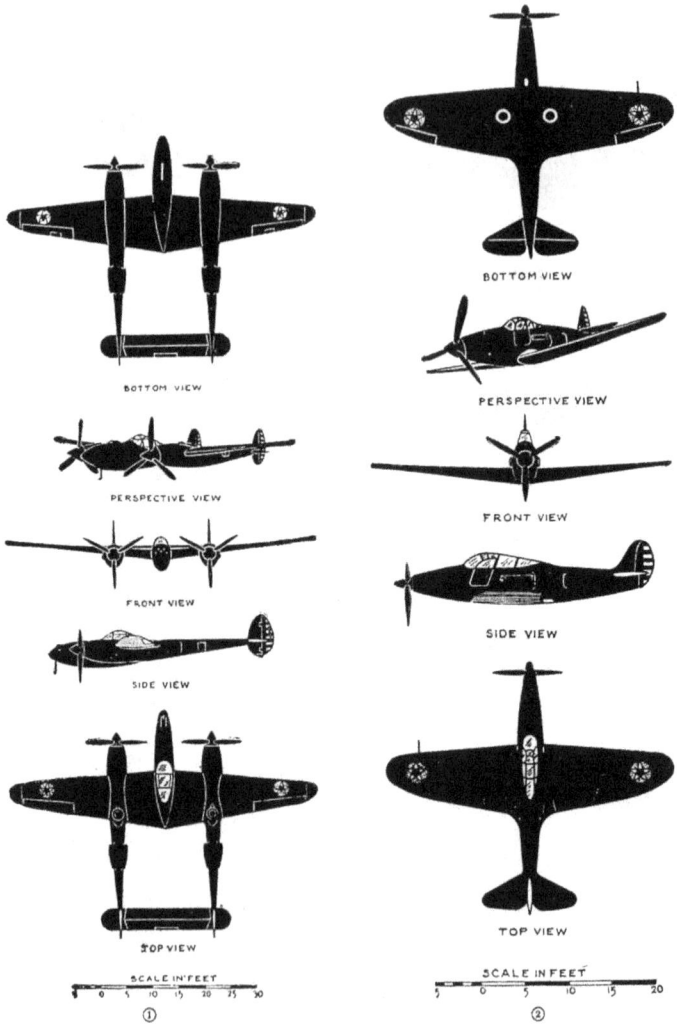

FIGURE 149.—Pursuit airplanes (interceptor).

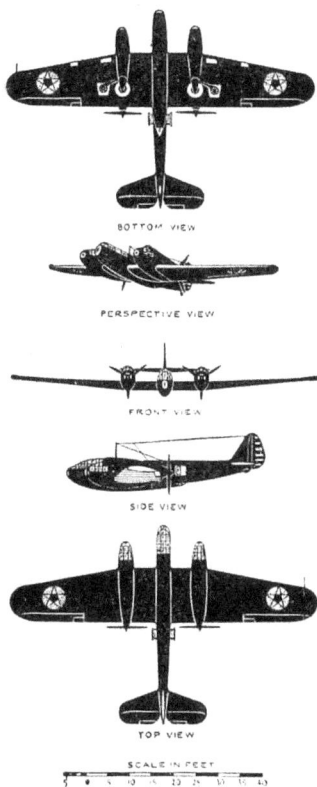

BOTTOM VIEW

PERSPECTIVE VIEW

FRONT VIEW

SIDE VIEW

TOP VIEW

SCALE IN FEET

FIGURE 150.—Multiplace fighter airplane.

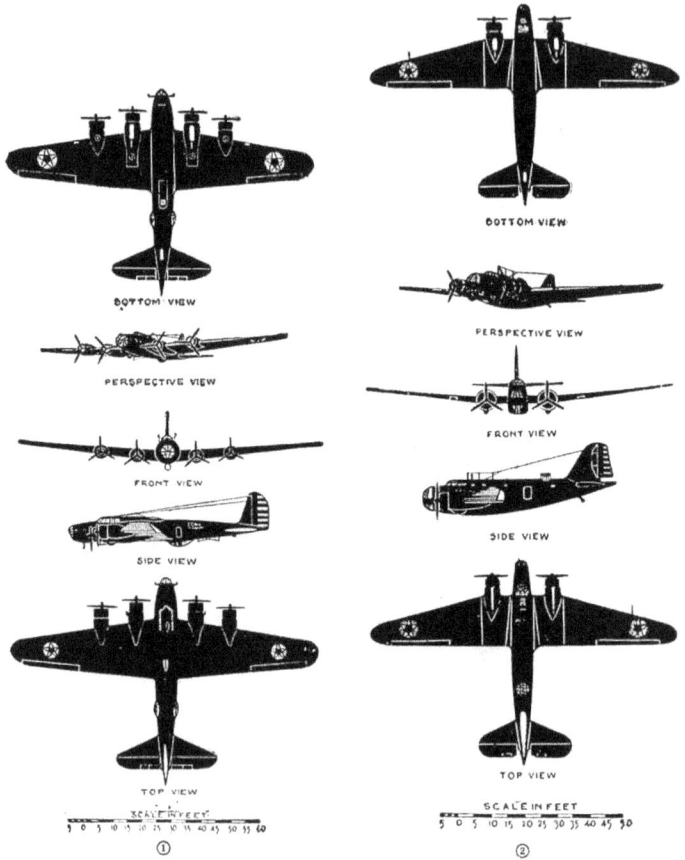

FIGURE 151.—Bombardment airplanes.

BOTTOM VIEW

PERSPECTIVE VIEW

FRONT VIEW

SIDE VIEW

TOP VIEW

SCALE IN FEET

③

BOTTOM VIEW

PERSPECTIVE VIEW

FRONT VIEW

SIDE VIEW

TOP VIEW

SCALE IN FEET

④

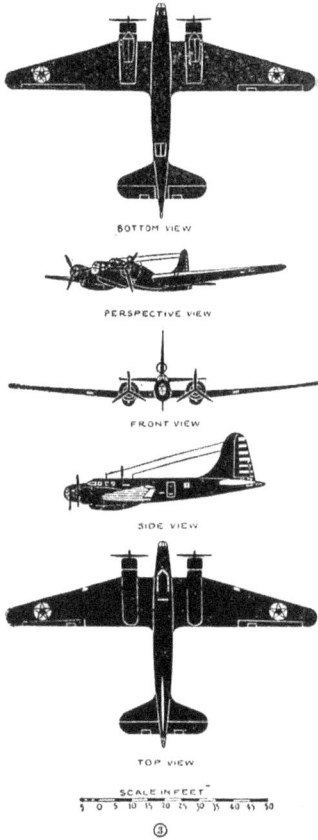

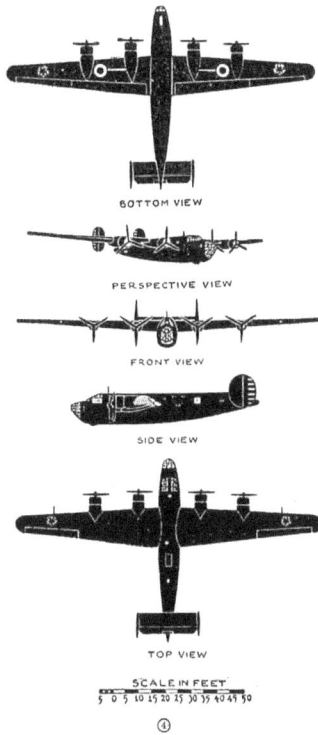

FIGURE 151.—Bombardment airplanes—Continued.

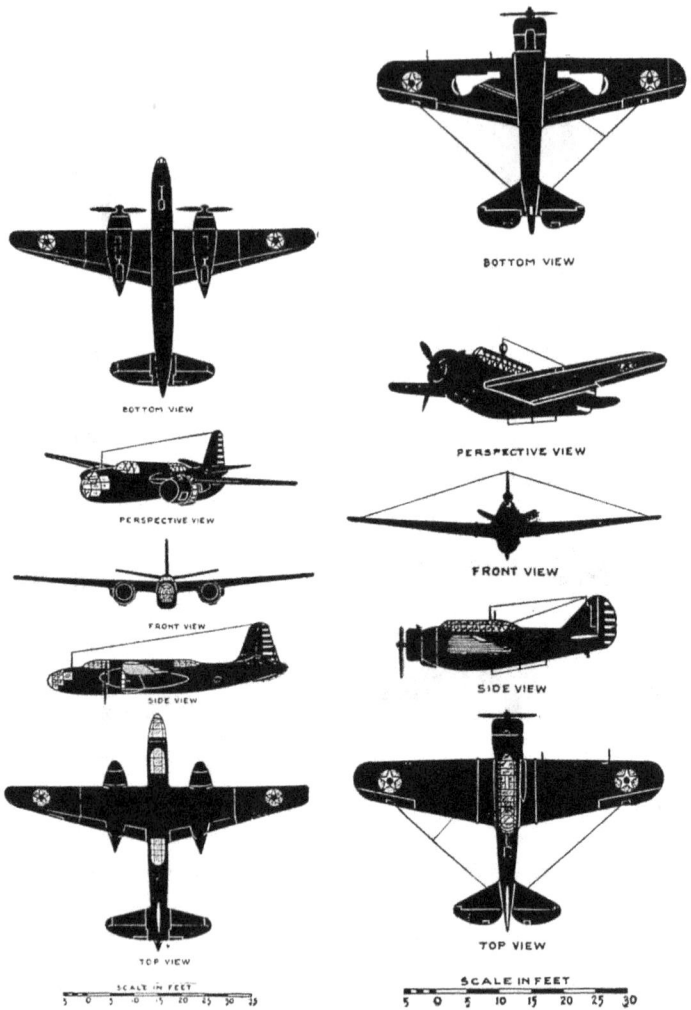

FIGURE 152.—Light bombardment (attack) airplane.   FIGURE 153.—Observation airplane.

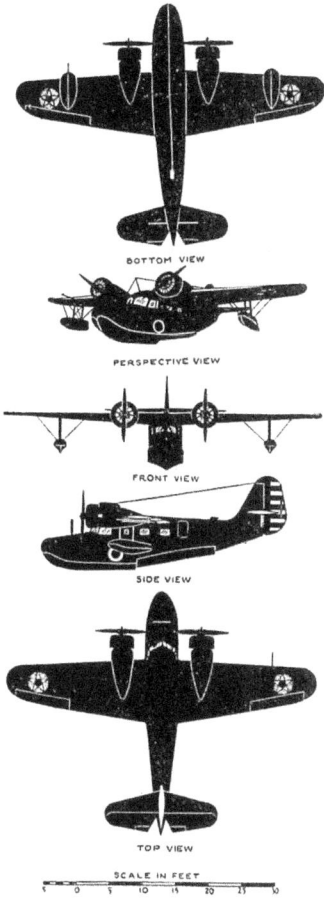

FIGURE 154.—Observation amphibian airplane.

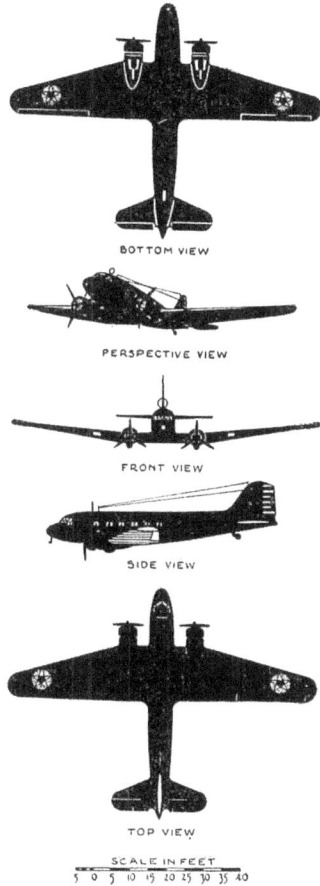

FIGURE 155.—Transport airplane.

INDEX

INDEX

INDEX

INDEX

INDEX

INDEX

[A. G. 062.11 (7–12–41).]

BY ORDER OF THE SECRETARY OF WAR:

> G. C. MARSHALL,
> *Chief of Staff.*

OFFICIAL:

> E. S. ADAMS,
> *Major General,*
> *The Adjutant General.*

DISTRIBUTION:

IBn and H 4 (3); IC 4 (AA Gn Btry (40); AA Auto W Btry (40); AA Hq Btry, Regt (30); AA Hq Btry, Bn (20); Repl Tng Cen—Gn and W Btrys (250); Repl Tng Cen—Hq Btry (125)).

(For explanation of symbols see FM 21–6.)

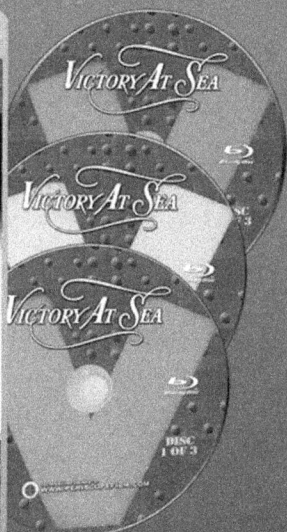

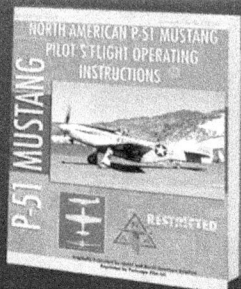

www.ingramcontent.com/pod-product-compliance
Lightning Source LLC
Chambersburg PA
CBHW060357200326
41518CB00009B/1173